Graduate Texts in Physics

Graduate Texts in Physics

Graduate Texts in Physics publishes core learning/teaching material for graduate- and advanced-level undergraduate courses on topics of current and emerging fields within physics, both pure and applied. These textbooks serve students at the MS- or PhD-level and their instructors as comprehensive sources of principles, definitions, derivations, experiments and applications (as relevant) for their mastery and teaching, respectively. International in scope and relevance, the textbooks correspond to course syllabi sufficiently to serve as required reading. Their didactic style, comprehensiveness and coverage of fundamental material also make them suitable as introductions or references for scientists entering, or requiring timely knowledge of, a research field.

More information about this series at http://www.springer.com/series/8431

Dietrich Stauffer · H. Eugene Stanley
Annick Lesne

From Newton to Mandelbrot

A Primer in Theoretical Physics

Third Edition

 Springer

Dietrich Stauffer
Institute for Theoretical Physics
Cologne University
Cologne
Germany

H. Eugene Stanley
Department of Physics—Center
   for Polymer Studies
Boston University
Boston, MT
USA

Annick Lesne
Laboratoire Physique Théorique de la
   Matière Condensée (LPTMC)
Université Pierre et Marie Curie
Paris
France

ISSN 1868-4513
Graduate Texts in Physics
ISBN 978-3-662-57151-4
DOI 10.1007/978-3-662-53685-8

ISSN 1868-4521   (electronic)

ISBN 978-3-662-53685-8   (eBook)

Printed on acid-free paper

This Springer imprint is published by Springer Nature
The registered company is Springer-Verlag GmbH Germany
The registered company address is: Heidelberger Platz 3, 14197 Berlin, Germany

Preface to the Third Edition

It's very simple: This English edition is an updated translation of our *Cours de Physique*, Springer, Paris 1999, which was an expanded French translation of Stauffer & Stanley, *From Newton to Mandelbrot*, Springer, New York 1990 and 1996, which in turn was an expanded translation of Stauffer, *Theoretische Physik*, Springer, Heidelberg 1989 (in German language). The first expansion was Chap. 5 on Fractals, the second expansion Chap. 6 on Dynamical Systems and Chaos. The present version omits the diskettes added to the earlier version.

Chapters 1–4 contain the standard material of courses in theoretical physics and are supposed to accompany lectures at the university; thus they are rather condensed. They are supposed to fill 1 year of teaching. Chapters 5 and 6, in contrast, are written less condensed since this material may not be part of standard lectures and thus could be studied without the help of a university teacher. An appendix on elementary particles lies somewhere in between: It could be a summary of a much more detailed course, or studied without such a course.

Cologne, Germany
Boston, Massachusetts
Paris, France
December 2015

Dietrich Stauffer
H. Eugene Stanley
Annick Lesne

Preface to the Second Edition

With increasing age, some authors gain more and more weight, scientifically and gravitationally, and so do their books. Thus a new section on elementary particle physics has been added. Its emphasis on computer simulation and phase transition connects it with the end of the Statistical Physics chapter. In contrast to the first four chapters, it does not lead to complicated exercises and may be more suited to self-study; thus it is put into an appendix. The first four chapters, thought to accompany a course or to summarize previous lectures, now also answer the many questions at the end of each chapter; instructors may get solutions of the more complicated problems by internet (stauffer@thp.uni-koeln.de). For the interested reader, we added to the four chapters recent literature references wherever modern research aspects are touched upon in the text.

Some computer programs for the fractals in Chap. 5 are included in the diskette that accompanies this book. More on the general subject of teaching fractals can be found in the book Fractals in Science, edited by H.E. Stanley, E.F. Taylor, and P.A. Trunfio (Springer, New York 1994, ISBN 0-387-94361-7 and 3-540-94361-7). The programs on the IBM diskette were constructed primarily by S.V. Buldyrev, F. Caserta, A. Chandra, K. Shakhnovich, and E.F. Taylor while those for the Macintosh diskette were written mainly by J. Blandey, S.V. Buldyrev, T. Mekonen, R.L. Selinger, P. Trunfio, and B. Volbright. We thank these individuals for their contribution, and also thank H. Rollnik, F.-W. Eicke, F.W. Hehl, E.W. Mielke, and J. Potvin for their help with the additions to the book. We hope readers who note further imperfections, or in any way wish to make constructive suggestions, will communicate their thoughts to the authors.

Cologne, Germany
Boston, Massachusetts
June 1995

Dietrich Stauffer
H. Eugene Stanley

Preface to the First Edition

This is not a book for theoretical physicists. Rather it is addressed to professionals from other disciplines, as well as to physics students who may wish to have in one slim volume a concise survey of the four traditional branches of theoretical physics. We have added a fifth chapter, which emphasizes the possible connections between basic physics and geometry. Thus we start with classical mechanics, where Isaac Newton was the dominating force, and end with fractal concepts, pioneered by Benoit Mandelbrot. Just as reading a review article should not replace the study of original research publications, so also perusing the present short volume should not replace systematic study of more comprehensive texts for those wishing a firmer grounding in theoretical physics.

The opening paragraphs of Chap. 5 benefitted from input by B. Jorgensen. We wish to thank G. Daccord for providing us with Plates 7 and 8, F. Family for Plates 1 and 15, A.D. Fowler for Plate 3, R. Lenormand for Plate 11, P. Meakin for Plate 14 as well as the cover illustration, J. Nittmann for Plate 13, U. Oxaal for Plate 10, A. Skjeltorp for Plates 4, 9 and 16, K.R. Sreenivasan for Plate 5, R.H.R. Stanley for Plate 2, and P. Trunfio for Plates 6 and 12. We also thank A. Armstrong, A. Coniglio, J. Hajdu, F.W. Hehl, K.W. Kehr, J. Kertesz, A. Margolina, R. Selinger, P. Trunfio, and D.E. Wolf as well as many students—particularly L. Jaeger—who offered their feedback at appropriate occasions and A. Armstrong for translating Chaps. 1–4 from the original German edition published by Springer.

Jülich and Boston   D. Stauffer
July 1990   H.E. Stanley

Contents

1	**Mechanics**		1
	1.1	Point Mechanics	1
		1.1.1 Basic Concepts of Mechanics and Kinematics	1
		1.1.2 Newton's Law of Motion	3
		1.1.3 Simple Applications of Newton's Law	6
		1.1.4 Harmonic Oscillator in One Dimension	13
	1.2	Mechanics of Point Mass Systems	17
		1.2.1 The Ten Laws of Conservation	17
		1.2.2 The Two-Body Problem	19
		1.2.3 Constraining Forces and d'Alembert's Principle	20
	1.3	Analytical Mechanics	24
		1.3.1 The Lagrange Function	24
		1.3.2 The Hamilton Function	26
		1.3.3 Harmonic Approximation for Small Oscillations	28
	1.4	Mechanics of Rigid Bodies	33
		1.4.1 Kinematics and Inertia Tensor	33
		1.4.2 Equations of Motion	36
	1.5	Continuum Mechanics	42
		1.5.1 Basic Concepts	42
		1.5.2 Stress, Strain and Hooke's Law	47
		1.5.3 Waves in Isotropic Continua	49
		1.5.4 Hydrodynamics	50
2	**Electricity and Magnetism**		61
	2.1	Vacuum Electrodynamics	61
		2.1.1 Steady Electric and Magnetic Fields	61
		2.1.2 Maxwell's Equations and Vector Potential	66
		2.1.3 Energy Density of the Field	68
		2.1.4 Electromagnetic Waves	68
		2.1.5 Fourier Transformation	69
		2.1.6 Inhomogeneous Wave Equation	70

	2.1.7	Applications	71
2.2		Electrodynamics in Matter	76
	2.2.1	Maxwell's Equations in Matter	76
	2.2.2	Properties of Matter	77
	2.2.3	Wave Equation in Matter	79
	2.2.4	Electrostatics at Surfaces	80
2.3		Theory of Relativity	83
	2.3.1	Lorentz Transformation	84
	2.3.2	Relativistic Electrodynamics	87
	2.3.3	Energy, Mass and Momentum	89

3 Quantum Mechanics 93
   3.1   Basic Concepts 93
       3.1.1   Introduction 93
       3.1.2   Mathematical Foundations 94
       3.1.3   Basic Axioms of Quantum Theory 96
       3.1.4   Operators 98
       3.1.5   Heisenberg's Uncertainty Principle 100
   3.2   Schrödinger's Equation 101
       3.2.1   The Basic Equation 101
       3.2.2   Penetration 102
       3.2.3   Tunnel Effect 104
       3.2.4   Quasi-classical WKB Approximation 105
       3.2.5   Free and Bound States in the Potential Well 106
       3.2.6   Harmonic Oscillators 107
   3.3   Angular Momentum and the Structure of the Atom 110
       3.3.1   Angular Momentum Operator 110
       3.3.2   Eigenfunctions of L^2 and L_z 111
       3.3.3   Hydrogen Atom 112
       3.3.4   Atomic Structure and the Periodic System 115
       3.3.5   Indistinguishability 116
       3.3.6   Exchange Reactions and Homopolar Binding 118
   3.4   Perturbation Theory and Scattering 120
       3.4.1   Steady Perturbation Theory 120
       3.4.2   Unsteady Perturbation Theory 122
       3.4.3   Scattering and Born's First Approximation 124

4 Statistical Physics 129
   4.1   Probability and Entropy 129
       4.1.1   Canonical Distribution 129
       4.1.2   Entropy, Axioms and Free Energy 132
   4.2   Thermodynamics of the Equilibrium 135
       4.2.1   Energy and Other Thermodynamic Potentials 135
       4.2.2   Thermodynamic Relations 138

4.2.3   Alternatives to the Canonical Probability
Distribution................................. 140

4.2.4   Efficiency and the Carnot Cycle................... 141

4.2.5   Phase Equilibrium and the Clausius-Clapeyron
Equation..................................... 143

4.2.6   Mass Action Law for Gases..................... 146

4.2.7   The Laws of Henry, Raoult and van't Hoff.......... 147

4.2.8   Joule-Thomson Effect.......................... 149

4.3   Statistical Mechanics of Ideal and Real Systems............ 150

4.3.1   Fermi and Bose Distributions..................... 150

4.3.2   Classical Limiting Case $\beta\mu \to -\infty$................ 152

4.3.3   Classical Equidistribution Law.................... 155

4.3.4   Ideal Fermi-Gas at Low Temperatures $\beta\mu \to +\infty$ 156

4.3.5   Ideal Bose-Gas at Low Temperatures $\beta\mu \to 0$........ 157

4.3.6   Vibrations................................... 160

4.3.7   Virial Expansion of Real Gases................... 161

4.3.8   Van der Waals' Equation........................ 162

4.3.9   Magnetism of Localised Spins.................... 164

4.3.10   Scaling Theory............................... 169

5   Fractals in Theoretical Physics............................. 189

5.1   Non-random Fractals............................... 190

5.2   Random Fractals: The Unbiased Random Walk............. 192

5.3   'A Single Length'................................. 194

5.3.1   The Concept of a Characteristic Length............. 194

5.3.2   Higher Dimensions............................ 195

5.3.3   Additional Lengths that Scale with $\sqrt{t}$............. 195

5.4   Functional Equations and Scaling: One Variable............ 196

5.5   Fractal Dimension of the Unbiased Random Walk........... 197

5.6   Universality Classes and Active Parameters................ 197

5.6.1   Biased Random Walk.......................... 197

5.6.2   Scaling of the Characteristic Length................ 198

5.7   Functional Equations and Scaling: Two Variables............ 200

5.8   Fractals and the Critical Dimension..................... 201

5.9   Fractal Aggregates................................. 207

5.10   Fractals in Nature................................. 210

6   Dynamical Systems and Chaos............................. 215

6.1   Basic Notions and Framework......................... 215

6.1.1   Phase Space................................. 215

6.1.2   Continuous-Time Dynamical Systems............... 216

6.1.3   Flows and Phase Portraits....................... 217

6.1.4   Insights from Dynamical Systems Theory............ 217

6.1.5   Some Examples.............................. 218

6.2 Fixed Points and Linear Stability Analysis 221
    6.2.1 Fixed Points and Stability Matrix 221
    6.2.2 Flow Linearization Theorem 222
    6.2.3 The Different Types of Fixed Points 223
    6.2.4 Constructing the Phase Portrait 225
    6.2.5 Application: Anharmonic Oscillators 227
    6.2.6 The Origin of Bifurcations 230
6.3 Attractors, Bifurcations and Normal Forms 231
    6.3.1 Attractors 231
    6.3.2 Conservative *Versus* Dissipative Systems 232
    6.3.3 The Different Types of Bifurcations 233
    6.3.4 Normal Forms and Structural Stability 235
6.4 Discrete-Time Dynamical Systems 235
    6.4.1 Discrete-Time Evolution Equations 235
    6.4.2 Linear Stability Analysis 236
    6.4.3 Attractors and Bifurcations 237
    6.4.4 Discretization by Poincaré Sections 237
6.5 Lyapunov Exponents and Deterministic Chaos 239
    6.5.1 Lyapunov Exponents 239
    6.5.2 Deterministic Chaos 240
    6.5.3 Ergodic Theory 242
6.6 Routes to Chaos 243
    6.6.1 Period Doubling and Subharmonic Cascade 243
    6.6.2 Intermittency 245
    6.6.3 Ruelle-Takens Scenario 246
    6.6.4 Hamiltonian Systems and KAM Theorem 246
6.7 Conclusion 248
6.8 Problems 249
6.9 Further Reading 250

Appendix A: Elementary Particles 253

Appendix B: Answers to Questions 263

Index 267

Chapter 1
Mechanics

Abstract From Kepler's laws to hydrodynamics via Lagrange and Hamilton functions this first chapter covers classical mechanics, i.e. without relativistic or quantum effects.

Theoretical physics is the first science to be expressed mathematically: the results of experiments should be predicted or interpreted by mathematical formulae. Mathematical logic, theoretical chemistry and theoretical biology arrived much later. Physics had been understood mathematically in Greece more than 2000 years earlier, for example the law of buoyancy announced by Archimedes—lacking *Twitter*—with *Eureka*! Theoretical Physics first really came into flower, however, with Kepler's laws and their explanation by Newton's laws of gravitation and motion. We also shall start from that point.

1.1 Point Mechanics

1.1.1 Basic Concepts of Mechanics and Kinematics

A point mass is a mass whose spatial dimension is negligibly small in comparison with the distances involved in the problem under consideration. Kepler's laws, for example, describe the earth as a point mass "circling" the sun. We know, of course, that the earth is not really a point, and geographers cannot treat it in their field of work as a point. Theoretical physicists, however, find this notion very convenient for describing approximately the motion of the planets: theoretical physics is the science of successful approximations. Biologists often have difficulties in accepting similarly drastic approximations in their field.

The motion of a point mass is described by a position vector r as a function of time t, where r consists of the three components (x, y, z) of a rectangular coordinate

© Springer-Verlag GmbH Germany 2017
D. Stauffer et al., *From Newton to Mandelbrot*, Graduate Texts in Physics,
DOI 10.1007/978-3-662-53685-8_1

system. (A boldface variable represents a vector. The same variable not in boldface represents the absolute magnitude of the vector, thus for example $r = |r|$). Its velocity v is the time derivative

$$v(t) = \frac{dr}{dt} = (\dot{x}, \dot{y}, \dot{z}),\tag{1.1}$$

where a dot over a variable indicates the derivative with respect to time t. The acceleration a is

$$a(t) = \frac{dv}{dt} = \frac{d^2r}{dt^2} = (\dot{v}_x, \dot{v}_y, \dot{v}_z),\tag{1.2}$$

the second derivative of the position vector with respect to time.

Galileo Galilei (1564–1642) discovered by experimentally dropping objects, presumably not from the Leaning Tower of Pisa, that all objects fall to the ground equally "fast", with the constant acceleration

$$a = g \quad \text{and } g = 9.81 \, m/s^2.\tag{1.3}$$

Nowadays this law can be conveniently "demonstrated" in the university lecture room by allowing a piece of chalk and a scrap of paper to drop simultaneously: both reach the floor at the same time ... don't they?

It will be observed that theoretical physics is often concerned with asymptotic limiting cases: (1.3) is valid only in the limiting case of vanishing friction, never fully achieved experimentally, just as good chemistry can be carried out only with "chemically pure" materials. Nature is so complex that natural scientists prefer to observe unnatural limiting cases, which are easier to understand. A realistic description of Nature must strive to combine the laws so obtained, in such a way that they describe the reality, and not the limiting cases.

The differential equation (1.3), $d^2r/dt^2 = (0, 0, -g)$ has for its solution the well known parabolic trajectory

$$r(t) = r_0 + v_0 t + (0, 0, -g)t^2/2,$$

where the z-axis is taken as usual to be the upward vertical. Here r_0 and v_0 are the position and the velocity initially (at $t = 0$); the number 1.3 is an equation number, denoted as 1.3 or eq. (1.3) etc. in other publications. It is more complicated to explain the motion of the planets around the sun; in 1609 and 1619 Johann Kepler accounted for the observations known at that time with the three Kepler laws:

1. Each planet moves on an ellipse with the sun at a focal point.
2. The radius vector r (from the sun to the planet) sweeps out equal areas in equal times.
3. The ratio (orbital period)^2/(major semi-axis)^3 has the same value for all planets in our solar system.

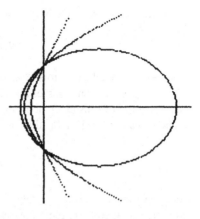

Ellipses are finite conic sections and hence differ from hyperbolae; the limiting case between ellipses and hyperbolae is the parabola. In polar coordinates (distance r, angle ϕ) we have

$$r = p/(1 + \varepsilon \cos \phi),$$

where $\varepsilon < 1$ is the eccentricity of the ellipse and the planetary orbit. (Circle $\varepsilon = 0$; parabola $\varepsilon = 1$; hyperbola $\varepsilon > 1$; see Fig. 1.1). Hyperbolic orbits are exhibited by comets; mathematically, Halley's Comet is not a comet *in this sense*, but a very eccentric planet.

It is remarkable, especially for modern science politicians, that from these laws of Kepler for the motion of remote planets, theoretical physics and Newton's law of motion resulted. Modern mechanics was derived, not from practical, "down to earth" research, but from a desire to understand the motion of the planets in order to produce better horoscopes. Kepler also occupied himself with snowflakes (see Chap. 5). That many of his contemporaries ignored Kepler's work, and that he did not always get his salary, places many of us today on a par with him, at least in this respect.

1.1.2 Newton's Law of Motion

Regardless of fundamental debates on how one defines "force" and "mass", we designate a reference system as an inertial system if a force-free body moves in a straight line with a steady velocity. We write the law of motion discovered by Isaac Newton (1642–1727) thus:

$$f = ma$$

force = mass × acceleration. $\qquad(1.4)$

For free fall we state Galileo's law (1.3) as

$$\text{weight} = mg. \tag{1.5}$$

Forces are added as vectors ("parallelogram of forces"), for two bodies we have action $= -$reaction, and masses are added arithmetically. So long as we do not need to take account of Einstein's theory of relativity, masses are independent of velocity. The *momentum* p is defined by $p = mv$, so that (1.4) may also be written as

$$f = \frac{dp}{dt}, \tag{1.6}$$

which remains valid even with relativity. The law action $= -$reaction then states for two mutually interacting point masses that

The sum of the momenta of the two masses remains constant.   (1.7)

It is crucial to these formulae that the force is proportional to the acceleration and not to the velocity. For thousands of years it was believed that there was a connection with the velocity, as is suggested by one's daily experience dominated by friction. For seventeenth century philosophers it was very difficult to accept that force-free bodies would continue to move with constant velocity; children of the space age have long been familiar with this idea.

It is not stipulated which of the many possible inertial systems is used: one can specify the origin of coordinates in one's office or in the government's Department of Education. Transformations from one inertial system to another ("Galileo transformations") are written mathematically as:

$$r' = \mathcal{R}r + v_0 t + r_0; \quad t' = t + t_0 \tag{1.8}$$

with arbitrary parameters v_0, r_0, t_0 (Fig. 1.2). Here $\mathcal{R}$ is a rotational matrix with three "degrees of freedom" (three angles of rotation); there are three degrees of freedom also in each of v_0 and r_0, and the tenth degree of freedom is t_0. Corresponding to these ten continuous variables in the general Galileo transformation we shall later find ten laws of conservation.

There are interesting effects if the system of reference is not an inertial system. For example we can consider a flat disk rotating (relative to the fixed stars) with an angular velocity $\omega = \omega(t)$ (Fig. 1.3). The radial forces then occurring are well known from rides on a carousel. Let the unit vector in the r direction be $e_r = r/|r|$, and the unit vector perpendicular to it in the direction of rotation be e_ϕ, where ϕ is the angle with the x-axis: $x = r \cos \phi$, $y = r \sin \phi$. The time derivative of e_r is ωe_ϕ, that of e_ϕ is $-\omega e_r$, with the *angular velocity* $\omega = d\phi/dt$. The velocity is

Fig. 1.2 Example of a
transformation (1.8) in
two-dimensional space

Fig. 1.3 Polar coordinates
(r, ϕ) on a flat disk rotating
with angular velocity ω,
viewed from above

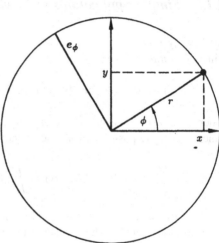

$$v = \mathrm{d}\,(r e_{\mathrm{r}})\,/\mathrm{d}t = e_{\mathrm{r}}\mathrm{d}r/\mathrm{d}t + r\omega e_{\phi}$$

according to the rule for the differentiation of a product Similarly for the acceleration a and the force f we have

$$\frac{f}{m} = a = \dot{v} = \left(\frac{\mathrm{d}^2 r}{\mathrm{d}t^2} - \omega^2 r\right) e_{\mathrm{r}} + (2\dot{r}\omega + r\dot{\omega})e_{\phi}. \tag{1.9}$$

Of the four terms on the right hand side the third is especially interesting. The first is "normal", the second is "centrifugal", the last occurs only if the angular velocity varies. In the case when, as at the north pole on the rotating earth, the angular velocity is constant, the last term disappears. The penultimate term in (1.9) refers to the Coriolis force and implies that in the northern hemisphere of the earth swiftly moving objects are deflected to the right, as observed with various phenomena on the rotating earth: Foucault's pendulum (1851), the precipitous right bank of the Volga, the direction of spin on the weather map for European depressions, Caribbean hurricanes and Pacific typhoons. For example, in an area of low pressure in the North Atlantic the air flows inwards; if the origin of our polar coordinates is taken at the centre of the depression (and for the sake of simplicity this is taken at the north pole), dr/dt is then negative, ω is constant, and the "deflection" of the wind observed from the rotating earth is always towards the right; at the south pole it is reversed. (If the observer is not at the north pole, ω has to be multiplied by $\sin \psi$, where ψ is the latitude: at the equator there is no Coriolis force.)

1.1.3   Simple Applications of Newton's Law

(a) Energy Law
Since $f = ma$ we have:

$$f\,dr/dt = m\,(dr/dt)\,(d^2r/dt^2) = d\left(mv^2/2\right)/dt = dT/dt,$$

where $T = mv^2/2$ is the *kinetic energy*. Accordingly the difference between the kinetic energy at position 1 (or time 1) and that at position 2 is given by

$$T\,(t_2) - T\,(t_1) = \int_1^2 f\,v\,dt = \int_1^2 f\,dr,$$

which corresponds to the mechanical work done on the point mass ("work = force times displacement"). (The product of two vectors such as f and v is here the scalar product, viz. $f_x v_x + f_y v_y + f_z v_z$. The multiplication point is omitted. The cross product of two vectors such as $f \times v$ comes later.) The power dT/dt ("power = work/time") is therefore equal to the product of force f and velocity v, as one appreciates above all on the motorway, but also in the study.

A three-dimensional force field $f(r)$ is called *conservative* if the above integral over $f\,dr$ between two fixed endpoints 1 and 2 is independent of the path followed from 1 to 2. The gravity force $f = mg$, for example, is conservative:

$$\int f\,dr = -mgh,$$

where the height h is independent of the path followed. Defining the *potential energy*

$$U(r) = -\int f\,dr$$

we then have:

> The force f is conservative if and only if a potential U exists such that
>
> $$f = -\text{grad}\, U = -\nabla U. \tag{1.10}$$

Here we usually have conservative forces to deal with and often neglect frictional forces, which are not conservative. If a point mass now moves from 1 to 2 in a conservative field of force, we have:

$$T_2 - T_1 = \int_1^2 f \, d\mathbf{r} = -(U_2 - U_1),$$

so that $T_1 + U_1 = T_2 + U_2$, i.e. $T + U = \text{const}$:

> The energy $T + U$ is constant in a conservative field of force. $\tag{1.11}$

Whoever can find an exception to this law of energy so central to our daily life can produce perpetual motion. We shall later introduce other forms of energy besides T and U, so that frictional losses ("heat") etc. can also be introduced into the energy law, allowing non-conservative forces also to be considered. Equation (1.11) shows mathematically that one can already predict important properties of the motion without having to calculate explicitly the entire course of the motion ("motion integrals").

(b) One-Dimensional Motion and the Pendulum
In one dimension all forces (depending on x only and thus ignoring friction) are automatically conservative, since there is only a unique path from one point to another point in a straight line. Accordingly $E = U(x) + mv^2/2$ is always constant, with $dU/dx = -f$ and arbitrary force $f(x)$. (Mathematicians should know that physicists pretend that all reasonable functions are always differentiable and integrable, and only now consider that known mathematical monsters such as "fractals" (see Chap. 5) also have physical meaning.) One can also see this directly:

$$dE/dt = (dU/dx)(dx/dt) + mv\, dv/dt = -f v + mva = 0.$$

Moreover we have $dt/dx = 1/v = [(E - U)2/m]^{-1/2}$, and hence

$$t = t(x) = \int \frac{dx}{\sqrt{(E - U(x))2/m}}. \tag{1.12}$$

Accordingly, to within an integration constant, the time is determined as a function of position x by a relatively simple integral. Many pocket calculators can already carry out integrations automatically at the push of a button. For harmonic oscillators,

Fig. 1.4 Periodic motion
between the points a and b,
when the energy E lies in the
trough of the potential $U(x)$

such as the small amplitude pendulum, or the weight oscillating up and down on
a spring, $U(x)$ is proportional to x^2, and this leads to sine and cosine oscillations
for $x(t)$, provided that the reader knows the integral of $(1 - x^2)^{-1/2}$. In general,
if the energy E results in a motion in a potential trough of the curve $U(x)$, there
is a periodic motion (Fig. 1.4), which however need not always be $\sin(\omega t)$. In the
anharmonic pendulum, for example, the restoring force is proportional to $\sin(x)$ (here
x is the angle), and the integral (1.12) leads to elliptic functions, which we do not
propose to pursue any further.

Notwithstanding the exact solution by (1.12), it is also useful to consider a computer program, with which one can solve $f = ma$ directly. Quite basically (I leave
better methods to the numerical mathematicians) one divides up the time into individual time steps Δt. If we know the position x at that time we can calculate the
force f and hence the acceleration $a = f/m$. The velocity v varies in the interval
Δt by $a \Delta t$, the position x by $v \Delta t$. We thus construct the command sequence of the
program PENDULUM, which is constantly to be repeated

```
calculate f(x)
replace v by v + (f/m)Δt
replace x by x + vΔt
return to calculation of f.
```

At the start we need an initial velocity v_0 and an initial position x_0. By suitable
choice of the unit of time the mass can be set equal to unity. Programmable pocket
calculators can be eminently suitable for executing this program. It is presented here
in the computer language BASIC for $f = -\sin x$. It is clear that programming can be
very easy; one should not be frightened by textbooks, where a page of programming
may be devoted merely to the input of the initial data.

```
PROGRAM PENDULUM
   10 x =0.0
   20 v =1.0
   30 dt=0.1
   40 f=-sin(x)
   50 v =v+f*dt
   60 x =x+v*dt
   70 print x,v
   80 goto 40
   90 end
```

In BASIC and FORTRAN

$$a = b + c \ (a := b + c); \ \text{(in PASCAL)}$$

signifies that the sum of b and c is to be stored at the place in store reserved for the variable a. The command

$$n = n + 1$$

is therefore not a sensational new mathematical discovery, but indicates that the variable n is to be increased by one from its previous value. By "goto" one commands the computer control to jump to the program line corresponding to the number indicated. In the above program the computer must be stopped by a command. In line 40 the appropriate force law is declared. It is of course still shorter if one simply replaces lines 40 and 50 by

$$40 \ v = v - \sin(x) * \mathrm{d}t.$$

About computer programming: The programs in this book are supposed to be *understood*, not to be merely used as black boxes. The language BASIC is used for them since that name suggests that the language is simple. FORTRAN is quite similar while C^{++} is different. Translating a BASIC program into your preferred programming language will help understanding it. For much longer programs than those in this book, structured programming with subroutines is recommended. Graphic commands have to be adjusted to the computer you use except if you print out all the numbers and then plot them by hand.

(c) Angular Momentum and Torque

The cross product $L = r \times p$ of position and momentum is the *angular momentum*, and $M = r \times f$ is the *torque*. Pedantic scientists might maintain that the cross product is not really a vector but an antisymmetric 3×3 matrix. We three-dimensional physicists can quite happily live with the pretence of handling L and M as vectors. As the analogue of $f = \mathrm{d}p/\mathrm{d}t$ we have

Fig. 1.5 The triangular area
swept out by the radius
vector r per unit time is a
half of the cross-product
$r \times v$. The upper picture is as
seen, looking along the axis.
The lower picture shows in
three dimensions the angle ϕ
and the vectors L and ω

$$M = \frac{\mathrm{d}L}{\mathrm{d}t}, \tag{1.13}$$

which can also be written as

$$M = r \times \dot{p} = \mathrm{d}(r \times p)/\mathrm{d}t - \dot{r} \times p = \dot{L},$$

and since the vector $\mathrm{d}r/\mathrm{d}t$ is parallel to the vector p, the cross product of the two
vectors vanishes. Geometrically $L/m = r \times v$ is twice the rate at which area is swept
out by the radius vector r (Fig. 1.5); the second law of Kepler therefore states that
the sun exerts no torque on the earth and therefore the angular momentum and the
rate at which area is swept out remain constant.

(d) Central Forces

Central Forces are those forces F which act in the direction of the radius vector r,
thus $F(r) = f(r)e_r$ with an arbitrary scalar function f of the vector r. Then the
torque $M = r \times F = (r \times r)f(r)/|r| = 0$:

Central forces exert no torque and leave the angular momentum unchanged.

$$\tag{1.14}$$

For all central forces the motion of the point mass lies in a plane normal to the
constant angular momentum L:

$$rL = r(r \times p) = p(r \times r) = 0$$

using the triple product rule

$$a(b \times c) = c(a \times b) = b(c \times a).$$

The calculation of the angular momentum in polar coordinates shows that for this motion ωr^2 remains constant: the nearer the point mass is to the centre of force, the faster it orbits round it *Question*: Does this mean that winter is always longer than summer?

(e) Isotropic Central Forces

Most central forces with which theoretical physicists have to deal are isotropic central forces. These are central forces in which the function $f(r)$ depends only on the magnitude $|r| = r$ and not on the direction: $F = f(r)e_r$. With

$$U(r) = -\int f(r)\mathrm{d}r$$

we then have $F = -\mathrm{grad}\,U$ and $f = -\mathrm{d}U/\mathrm{d}r$: the potential energy U also depends only on the distance r. Important examples are:

$U \sim 1/r$, so $f \sim 1/r^2$:	gravitation, Coulomb's law;
$U \sim \exp(-r/\xi)/r$:	Yukawa potential; screened Coulomb potential;
$U = \infty$ for $r < a$, $U = 0$ for $r > a$:	hard spheres (billiard balls);
$U = \infty$, $-U_0$ and 0 for $r < a$, $a < r < b$ and $r > b$:	
	spheres with potential well;
$U \sim (a/r)^{12} - (a/r)^6$:	Lennard-Jones or "6–12" potential;
$U \sim r^2$:	harmonic oscillator.

(Here $\sim$ is the symbol for proportionality, also denoted by the symbol $\propto$.)

For the computer simulation of real gases such as argon the Lennard-Jones potential is the most important: one places 10^6 such point masses in a computer and moves each according to force = mass × acceleration, where the force is the sum of the Lennard-Jones forces from the neighbouring particles. This method is called "molecular dynamics" and uses a lot of computer time.[1]

Since there is always a potential energy U, isotropic central forces are always conservative. If one constructs any apparatus in which only gravity and electrical forces occur, then the energy $E = U + T$ is necessarily constant. In a manner similar to the one-dimensional case the equation of motion can here be solved exactly, by resolving the velocity v into a component $\mathrm{d}r/\mathrm{d}t$ in the r-direction and a component $r\mathrm{d}\phi/\mathrm{d}t = r\omega$ perpendicular thereto and applying $L = m\omega r^2$:

$$E = U + T = U + \frac{1}{2}(mv^2)$$

$$= U + \frac{1}{2}m\left[(\mathrm{d}r/\mathrm{d}t)^2 + r^2\omega^2\right] = U + \frac{1}{2}m\left[(\mathrm{d}r/\mathrm{d}t)^2 + L^2/m^2r^2\right].$$

---

[1] W.G. Hoover, Computational Statistical Mechanics (Elsevier, Amsterdam 1991).

(In order to economise on parentheses, physicists often write a/bc for the fraction $a/(bc)$). Accordingly, with $U_{\text{eff}} = U + L^2/2mr^2$, we have:

$$\frac{dr}{dt} = \sqrt{2(E - U_{\text{eff}}/m)}, \quad t = \int \frac{dr}{\sqrt{2(E - U_{\text{eff}}/m)}}. \tag{1.15}$$

By defining the effective potential U_{eff} we can thus reduce the problem to the same form as in one dimension (1.12). However, we now want to calculate also the angle $\phi(t)$, using

$$L = mr^2\omega = mr^2\frac{d\phi}{dr}\frac{dr}{dt} : \quad \frac{d\phi}{dr} = \frac{L}{mr^2\sqrt{2(E - U_{\text{eff}})/m}}. \tag{1.16}$$

Integration of this yields $\phi(r)$ and everything is solved.

(f) Motion in a Gravitational Field

Two masses M and m separated by a distance r attract each other according to Newton's law of gravity

$$U = -GMm/r \text{ and } f = -GMm/r^2, \tag{1.17}$$

where G the gravitational constant is equal to 6.67×10^{-8} in cgs units. (The old familiar centimetre-gram-second units such as ergs and dynes are still in widespread use in theoretical physics; 1 dyne $= 10^{-5}$newton $= 1\,\text{g cm/s}^2$; 1 erg $= 10^{-7}$joule or watt $-$ second $= 1\,\text{g cm}^2/\text{s}^2$.) Unlike mutually repulsive electrical charges, mutually repulsive masses have so far not been discovered. For planets M is the mass of the sun and m is the mass of the planet.

Since

$$\int (1 - x^2)^{-1/2}dx = -\arccos x$$

integration of (1.16) leads to the result

$$r = p/(1 + \varepsilon \cos \phi)$$

corresponding to Kepler's ellipse law of Sect. 1.1.1 with the parameter $p = L^2/GMm^2$ and the eccentricity $\varepsilon = (1 + 2Ep/GMm)^{1/2}$. For large energies $\varepsilon > 1$ and we obtain a hyperbola (comet) instead of an ellipse ($\varepsilon < 1$). Kepler's second law states, as mentioned above, the conservation of angular momentum, a necessary consequence of isotropic central forces such as gravitation. The third law, moreover, states that

$$\frac{(\text{period})^2}{(\text{major semi} - \text{axis})^3} = \frac{4\pi^2}{GM}. \tag{1.18}$$

(The derivation can be made specially simple by using circles instead of ellipses and then setting the radial force $m\omega^2 r$ equal to the gravitational force GMm/r^2 : period $= 2\pi/\omega$.)

The computer simulation also makes it possible to allow hypothetical deviations from the law of gravitation, e.g., $U \sim 1/r^2$ instead of $1/r$. The computer simulation shows that there are then no closed orbits at all. The BASIC program PLANET illustrates only the correct law, and with the inputs 0.5, 0, 0.01 leads to a nice ellipse, especially if one augments the program with the graphic routine appropriate for the computer in use. In contrast to our first program, we are here dealing with two dimensions, using x and y for the position and v_x and v_y for the velocity; the force also must be resolved into x- and y-components: $f_x = xf/r$, $f_y = yf/r$. ("Input" indicates that one should key in the numbers for the start of the calculation, and "sqr" is the square root.) For an artificial law of gravitation with $U \sim 1/r^2$ one has only to replace the root "sqr(r2)" in line 50 by its argument "r2"; the graphics will then show that nothing works so well any more.

```
PROGRAM PLANET
    10 input "vx,vy,dt ="; vx,vy,dt
    20 x =0.0
    30 y =1.0
    40 r2=x*x+y*y
    50 r3=dt/(r2*sqr(r2))
    60 vx=vx-x*r3
    70 vy=vy-y*r3
    80 x =x+dt*vx
    90 y =y+dt*vy
    100 print x,y
    110 goto 40
    120 end
```

1.1.4 Harmonic Oscillator in One Dimension

The harmonic oscillator appears as a continuous thread through theoretical physics and is defined in mechanics by

$$T = mv^2/2, \ U = Kx^2/2, \ E = T + U = p^2/2m + Kx^2/2. \tag{1.19}$$

For example, a weight hanging on a spring moves in this way, provided that the displacement x is not too great, so that the restoring force is proportional to the displacement.

(a) Without Friction

The calculation of the integral (1.12) with $\omega^2 = K/m$ gives the solution

$$x = x_0 \cos(\omega t + \text{const.}),$$

which one can however obtain directly: it follows from (1.4) that

$$m\frac{d^2x}{dt^2} + Kx = 0, \tag{1.20}$$

and the sine or the cosine is the solution of this differential equation. The potential energy oscillates in proportion to the square of the cosine, the kinetic energy in proportion to the square of the sine; since $\cos^2\psi + \sin^2\psi = 1$ the total energy $E = U + T$ is constant, as it must be.

In electrodynamics we shall come across light waves, where the electric and magnetic field energies oscillate. In quantum mechanics we shall solve (1.19) by the Schrödinger equation and show that the position x and the momentum p cannot both be exactly equal to zero ("Heisenberg's Uncertainty Principle"). In statistical physics we shall calculate the contribution of vibrations to the specific heat, for application perhaps in solid state physics ("Debye Theory"). Harmonic oscillations are also well known in technology, for example as oscillating electrical currents in a coil ($\approx$ kinetic energy) and a condenser ($\approx$ potential energy), with friction corresponding to the electrical resistance ("Ohm's Law").

(b) With Friction

In theoretical physics (not necessarily in reality) frictional forces are usually proportional to the velocity. We therefore assume a frictional force $-R dx/dt$,

$$m d^2x/dt^2 + R dx/dt + Kx = 0.$$

This differential equation (of the second order) is linear, i.e. it involves no powers of x, and has constant coefficients, i.e. m, R and K are independent of t. Such differential equations can generally be solved by complex exponential functions $\exp(i\phi) = \cos\phi + i\sin\phi$, of which one eventually takes the real part. In this sense we try the solution

$$x = a\, e^{i\omega t} \rightarrow dx/dt = i\omega x \rightarrow d^2x/dt^2 = -\omega^2 x$$

and try to find the complex numbers a and ω. For the case without friction (1.20) is quite simple:

$$-m\omega^2 x + Kx = 0, \text{ or } \omega^2 = K/m.$$

With friction we now obtain

$$-m\omega^2 x + i\omega Rx + Kx = 0.$$

This quadratic equation has the solution

$$\omega = \mathrm{i}R/2m \pm \sqrt{K/m - R^2/4m^2}.$$

If we resolve ω into its real part Ω and its imaginary part $1/r$, $\omega = \Omega + \mathrm{i}/r$, we obtain

$$x = a\mathrm{e}^{\mathrm{i}\Omega t}\mathrm{e}^{-t/r}.$$

Quite generally, with a complex frequency $\omega = \Omega + \mathrm{i}/\tau$ the real part Ω corresponds to a cosine oscillation and the imaginary part to a damping with a decay time r. In the above expression, if we set $a = 1$ for simplicity, the real part is

$$x = \cos(\Omega t)\mathrm{e}^{-t/r}. \tag{1.21}$$

Similarly other linear differential equations of n-th order with constant coefficients can be reduced to a normal equation with powers up to ω^n. The imaginary and complex numbers, with $\mathrm{i}^2 = -1$, existing originally only in the imagination of mathematicians, have thus become a useful tool in practical physics.

So what does the above result mean? If $4K/m > R^2/m^2$, then the square root is real and equal to Ω, and $1/\tau = R/2m$. Then (1.21) describes a damped oscillation. If on the other hand $4K/m < R^2/m^2$, then the square root is purely imaginary, ω no longer has a real part Ω, and we have an overdamped, purely exponentially decaying motion. The "aperiodic limiting case" $4Km = R^2$ involves a further mathematical difficulty ("degeneracy") which we willingly leave to the shock absorption engineers. Figure 1.6 shows two examples.

(c) Resonance

We shall discuss resonance effects when a damped harmonic oscillator moves under the influence of a periodic external force. "As everyone knows", resonance presupposes that the oscillator and the force have about the same oscillation frequency.

We again use complex numbers in the calculation; the external force, which obeys a sine or cosine law, is accordingly expressed as a complex oscillation $f\exp(\mathrm{i}\omega t)$, and not as proportional to $\cos(\omega t)$. Then the inhomogeneous differential equation

Fig. 1.6 Displacement x as a function of time t for $K = 1, m = 1$ and $R = 1$ (oscillation) and $R = 4$ (damping). In the first case the decay period τ is marked becomes:

$$md^2x/dt^2 + Rdx/dt + Kx = fe^{i\omega t}.$$

The trial solution $x = a\exp(i\omega t)$ leads again to the algebraic equation

$$-m\omega^2 a + Ri\omega a + Ka = f;$$

the factor $\exp(i\omega t)$ has dropped out, retrospectively justifying the trial solution. It is clearer if we put $\omega_0^2 = K/m$ and $1/\tau = R/m$, since ω_0 is the eigenfrequency of the oscillator without the external force, and τ is its decay time. The above equation can be solved quite simply:

$$a = (f/m)/\left(\omega_0^2 - \omega^2 + i\omega/\tau\right).$$

This amplitude a is a complex number, $a = |a|\exp(-i\psi)$, where the "phase" ψ represents the angle by which the oscillation x lags behind the force f. The modulus, given by $|a|^2 = (\text{Re}\,a)^2 + (\text{Im}\,a)^2$, is of greater interest:

$$|a| = \frac{(f/m)}{\sqrt{(\omega_0^2 - \omega^2)^2 + \omega^2/\tau^2}}. \tag{1.22}$$

This function $|a|$ of ω is an even function, i.e. its value is independent of the sign of ω, so we can now assume $\omega \geq 0$. If the friction is small, so that τ is large, then this function looks something like Fig. 1.7: a relatively narrow peak has its maximum in the neighbourhood of $\omega = \omega_0$, the width of this maximum being of the order $1/\tau$. (Experts will know that it is not the amplitude, but the energy loss through friction, that is maximal when $\omega = \omega_0$; for weak damping the difference is unimportant.) Similar phenomena often occur in physics: the eigenfrequency is given approximately by the position of the resonance maximum, the reciprocal of the decay time by the width.

Fig. 1.7 Representation of the resonance function (1.22) for small damping. The maximum lies close to the eigenfrequency ω_0, the width is given by $1/\tau$

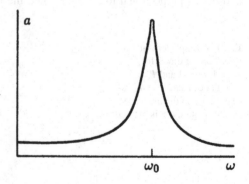

When $\omega = \omega_0$ then $|a| = (f/m)/(\omega_0/\tau) = f\tau/(Km)^{1/2}$. The smaller the damping, the longer is the decay time, and so the higher is the maximum near $\omega = \omega_0$. In the limiting case of infinitely small damping, $\tau = \infty$, there is an infinitely high and infinitely sharp maximum, and a resonance experiment is then impossible in practice: one would have to hit the correct frequency ω_0 exactly. It is therefore realistic to have only a very weak damping, and the effect of radio tuning is well known. One has to set the frequency approximately right in order to obtain a perceptible amplification. The smaller is the damping, the more exactly one has to hit the required frequency. One who finds radio programs too boring might well study the film of a particularly elegant road bridge (Tacoma Narrows, U.S.A.), which collapsed decades ago, as the wind produced oscillation frequencies which coincided with the eigenfrequencies of the torsional oscillations of the bridge. (A historian compared this instability with the international situation leading to the First World War; another compared Newton's law of gravity with the difficulty of keeping even far-away parts of an empire under control of the central government.)

1.2 Mechanics of Point Mass Systems

Up to this point we have considered a single point mass in a steady force field; in this section we pass on to several moving point masses, exerting forces upon each other. We shall find a complete solution for two such point masses; for more than two we restrict ourselves to general conservation laws.

1.2.1 The Ten Laws of Conservation

(a) Assumptions
Let there be N point masses with masses m_i, $i = 1, 2, ..., N$, which exert the forces $F_{ik} = -F_{ki}$ mutually between pairs. All these forces are isotropic central forces, i.e. mass k exerts on mass i the force

$$F_{ki} = f_{ki}\,(r_{ki})\,r_{ki}/\,|r_{ki}| = f_{ki}\,(r_{ki})\,e_{ki}$$

with $r_{ki} = r_k - r_i$. For convenience we define $f_{ii} = 0$ and then have to solve the following equations of motion:

$$m_i d^2 r_i/dt^2 = \sum_k f_{ki} e_{ki}.$$

(b) Energy Law
Let the kinetic energy $T = \Sigma_i T_i = \Sigma_i m_i v_i^2/2$ be the sum of all the particle energies T_i, the potential energy U the double sum $\Sigma_i \Sigma_k U_{ik}/2$ of all the two-particle potentials

U_{ik} and let there be no explicit dependence on time. Then the energy conservation law is:

$$\text{The energy } E = U + T \text{ is constant in time.} \tag{1.23}$$

Proof

$$dT/dt = \sum_i m_i v_i \dot{v}_i = \sum_{ik} f_{ki} e_{ki} v_i = \sum_{ik} (f_{ki} e_{ki} v_i + f_{ik} e_{ik} v_k)/2$$

$$= \sum_{ik} e_{ki} (v_i - v_k) f_{ki}/2 = -\sum_{ik} e_{ik} \dot{r}_{ki} f_{ki}/2$$

$$= -\sum_{ik} (\partial U_{ki}/\partial r_{ki}) \dot{r}_{ki}/2 = -dU/dt,$$

where $f_{ki} = f_{ik}$ and $e_{ki} = -e_{ik}$ has been used. Energy conservation with its associated problems is therefore based here on the chain-rule of differentiation and the exchange of indices in the double sums. (The partial derivative $\partial f/\partial x$ of a function $f(x, y, z, ...)$ is the derivative at constant $y, z, ...$)

(c) Momentum Law
The total momentum P, hence the sum $\Sigma_i p_i$ of the individual momenta, is likewise constant:

$$dP/dt = \sum_i \sum_k e_{ki} f_{ki} = \sum_{ik} (e_{ki} + e_{ik}) f_{ki}/2 = 0,$$

$$\text{The momentum } P \text{ is constant in time.} \tag{1.24}$$

(d) Law of Centre of Mass
$R = \Sigma_i m_i r_i / \Sigma_i m_i$ is the *centre of mass*, and $M = \Sigma_i m_i$ is the total mass. Since both P and M are constant, the velocity V of the centre of mass is also constant, because $P = \Sigma_i m_i v_i = M dR/dt = MV$. Hence

$$R = R_0 + V t. \tag{1.25}$$

It is often appropriate to choose the "centre of mass system" as the system of reference, in which the centre of mass lies always at the origin: $V = R_0 = 0$ (Fig. 1.8).

Fig. 1.8 Divorce in outer space: the two point masses fly asunder, but their centre of mass remains fixed

(e) Angular Momentum Law

For the constancy of the total angular momentum $L = \Sigma_i L_i$ we use

$$r_i \times F_{ki} + r_k \times F_{ik} = r_{ik} \times F_{ik} = 0.$$

Hence one can show that

> The angular momentum L is constant in time. (1.26)

Altogether we have here found ten conservation laws, since the constants P, V and L each have three components; E has only one component. Later we shall explain how these ten conservation laws are associated with ten "invariants"; for example, the total angular momentum is constant since no external torque is present and since therefore the total potential is invariant (unchanged) in a rotation of the whole system through a specified angle.

1.2.2 The Two-Body Problem

Systems with two point masses have simple and exact solutions. Let there be two point masses with isotropic central forces. We have to reconcile 12 unknowns (r and v for each of the two particles), the ten conservation quantities given above and Newton's laws of motion for the two particles. The problem should therefore be solvable. We use the centre of mass reference system already recommended above.

In this system we have $r_1 = -(m_2/m_1)r_2$, so that $r = r_1 - r_2$ and $e_r = e_{21}$ lie in the direction to r_1 from r_2. We therefore have:

$$d^2 r/dt^2 = e_{21} f_{21}/m_1 - e_{12} f_{21}/m_2 = e_r f_{21}\left(\frac{1}{m_1} + \frac{1}{m_2}\right) = e_r f(r)/\mu.$$

Accordingly Newton's law of motion is valid for the difference vector r, with an effective or reduced mass μ:

$$\mu \frac{d^2 r}{dt^2} = e_r f(r) \quad \text{with} \quad \mu = \frac{m_1 m_2}{m_1 + m_2}. \tag{1.27}$$

The problem of the two point masses has therefore been successfully reduced to the already solved problem of a single point mass.

In the motion of the earth around the sun the latter does not, of course, stand still but also moves, despite Galileo, around the combined centre of mass of the sun-earth system, which however lies inside the surface of the sun. The earth, like the sun, rotates on an ellipse, whose focal point lies at the centre of mass. Kepler's second law also applies to this centre of mass, both for the earth and for the sun. In Kepler's third law, where different planets are compared, there is now introduced a correction factor $m/\mu = (M + m)/M$, which is close to unity if the planetary mass m is very much smaller than the solar mass M. This correction factor was predicted theoretically and confirmed by more exact observations: a fine, if also rare, example of successful collaboration between theory and experiment.

In reality this is of course still inaccurate, since many planets are simultaneously orbiting round the sun and all of them are exerting forces upon each other. This many-body problem can be simulated numerically on the computer for many millions of years; but eventually the errors can become very large because the initial positions and velocities are not known exactly (and also because of the limited accuracy of the computer and the algorithm). Physicists call it "chaos" (see Chap. 5) when small errors can increase exponentially and make the eventual behaviour of the system unpredictable.[2] If the exhausted reader therefore lets this book fall to the ground, that tiny tremor will later cause so great an effect in the planetary system (supposing that this system is chaotic) that the decay of the planets' accustomed orbits will thereby be affected (positively or negatively). This will, however, not take place before your next exams!

1.2.3 Constraining Forces and d'Alembert's Principle

In reality the earth is not an inertial system, even if we "ignore" the sun, because of the gravitational force with which the earth attracts all masses. Billiard balls on a smooth table nevertheless represent approximately free masses, since they are constrained to move on the horizontal table. The force exerted by the smooth table on the balls exactly balances the force of gravity. This is a special case of the general conditions of constraint now to be considered, in which the point masses are kept within certain restrictive conditions (in this case on the table).

(a) Restrictive Conditions
We shall deal only with holonomic-scleronomic restrictive conditions, which are given by a condition $f(x, y, z) = 0$. Thus the motion on a smooth table of height $z = h$ means that the condition $0 = f(x, y, z) = z - h$ is fulfilled, whereas $0 = f =$

---

[2]H.G. Schuster, Deterministic Chaos (Physik Verlag, Weinheim, second edition 1989); M. Schroeder, Fractals, Chaos, Power Laws (Freeman, New York 1991); J. Guckenheimer and P. Holmes, Nonlinear Oscillations, Dynamical Systems, and Bifurcation of Vector Fields (Springer, Berlin, Heidelberg 1983).

$z \cdot \tan(\alpha) - x$ represents a sloping plane with inclination angle α. In general $f = 0$ indicates a surface, whereas the simultaneous fulfillment of two conditions $f_1 = 0$ and $f_2 = 0$ characterises a line (intersection of two surfaces).

The opposite of scleronomic (fixed) conditions are rheonomic (fluid) conditions of the type $f(x, y, z, t) = 0$. Non-holonomic conditions on the other hand can only be represented differentially: $0 = a \cdot dx + b \cdot dy + c \cdot dz + e \cdot dt$, and not by some function $f = 0$. Railways run on fixed tracks, whose course can be described by an appropriate function $f(x, y, z, t) = 0$: holonomic. Cars are, in contrast, non-holonomic: the motion $d\boldsymbol{r}$ follows the direction of the wheels, which one can steer. So in parking, for example, one can alter the y-coordinate at will, for a specified x-coordinate, by shuffling backwards and forwards in the x-direction (and more or less skillful steering). This shunting is not describable holonomically by $f(x, y) = 0$. The car is rheonomic, because one turns the steering-wheel, the railway is scleronomic.

(b) Constraining Forces

Those forces which hold a point mass on a prescribed path by the (holonomic-scleronomic) restrictive conditions are called *constraining forces* $\boldsymbol{Z}$. The billiard balls are held on the horizontal table by the constraining forces which the table exerts on them and which sustain the weight. The other forces, which are not constraining forces, are called *imposed forces* $\boldsymbol{F}$. We accordingly have: $m d^2 \boldsymbol{r}/dt^2 = \boldsymbol{F} + \boldsymbol{Z}$, the constraining forces act perpendicularly to the surface (or curve) on which the point mass has to move, and only the imposed forces can cause accelerations along the path of the point masses.

Mathematically the gradient grad $f = \nabla f = (\partial f/\partial x, \partial f/\partial y, \partial f/\partial z)$ is perpendicular to the surface defined by $f(x, y, z) = 0$. The constraining force is therefore parallel to grad f

$$\boldsymbol{Z} = \lambda \nabla f \qquad \text{(one condition)}$$
$$\boldsymbol{Z} = \lambda_1 \nabla f_1 + \lambda_2 \nabla f_2 \quad \text{(two conditions)},$$

with $\lambda = \lambda(\boldsymbol{r}, t)$. We accordingly have the Lagrange equations of the first kind for one and two conditions, respectively:

$$m \frac{d^2 \boldsymbol{r}}{dt^2} = \boldsymbol{F} + \lambda \nabla f, \quad m \frac{d^2 \boldsymbol{r}}{dt^2} = \boldsymbol{F} + \lambda_1 \nabla f_1 + \lambda_2 \nabla f_2. \qquad (1.28)$$

after Joseph Louis Comte de Lagrange (born in 1736 as Guiseppe Luigi Lagrangia in Turin, working also in Berlin.)

In practice one can solve this equation by resolving the imposed force $\boldsymbol{F}$ into one component $\boldsymbol{F}_t$ tangential and another component $\boldsymbol{F}_n$ normal (perpendicular) to the surface or curve of the restrictive condition: $\boldsymbol{F} = \boldsymbol{F}_t + \boldsymbol{F}_n, \boldsymbol{Z} = \boldsymbol{Z}_n + 0 = -\boldsymbol{F}_n$. Something on an inclined plane is treated quite simply in this way, as you learn at school; we use it instead to treat the pendulum (Fig. 1.9):

Fig. 1.9 Constraining force
and imposed gravity force in
a pendulum, with resolution
into normal and tangential
components

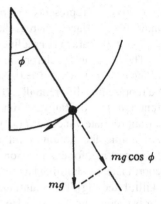

A mass m hangs on a string of length l; the string is tied to the origin of coordinates. Mathematically this is signified by the restriction $0 = f(r) = |r| - l$; hence grad $f = e_r$ and $Z = \lambda e_r$: the string force acts along the string. The resolution of the imposed gravity force $F = mg$ into tangential component $F_t = -mg \sin \phi$ and normal component $F_n = -mg \cos \phi$ (ϕ = angular displacement) gives $ml d^2\phi/dt^2 = ma_t = F_t = -mg \sin \phi$. The mass cancels out (since gravity mass = inertial mass), and there remains only the pendulum equation already treated in Sect. 1.1.3b. Monsieur Lagrange has therefore told us nothing new, but we have demonstrated with this familiar example that the formalism gives the correct result.

(c) Virtual Displacement and d'Alembert's Principle

We define a *virtual displacement* as an infinitely small displacement of the point mass such that the restrictive conditions are not violated. ("Infinitely small" in the sense of the differential equation: in $f'(x) = dy/dx$, dy is the variation in the function $y = f(x)$ caused by an infinitely small variation dx.) With an inclined plane this virtual displacement is therefore a displacement along the plane, without leaving it.

A virtual displacement δr accordingly occurs along the surface or the curve representing the restrictive conditions and is therefore perpendicular to the constraining force Z. Constraining forces therefore do no work: $Z\delta r = 0$, as is known from curriculum reform. Since $-Z = F - ma$ we have:

$$\left(F - m d^2 r/dt^2\right) r = 0; \tag{1.29}$$

$$\text{in equilibrium:} \qquad F\delta r = 0; \tag{1.30}$$

$$\text{if } F \text{ is conservative:} \quad \delta U = \nabla U \delta r = 0. \tag{1.31}$$

One generalises this principle to a system of N point masses m_i ($i = 1, 2, ..., N$) with ρ restrictive conditions $f_\mu = 0$, ($\mu = 1, 2, ..., \rho$), so we have

$$\text{Lagrange 1st kind:} \quad m_i d^2 r_i/dt^2 = F_i + \sum_\mu \lambda_\mu \nabla_i f_\mu(r_1, ..., r_N); \tag{1.32}$$

Fig. 1.10 Atwood's
Machine or: How the
theoretical physicist presents
an experimental apparatus

$$\text{d'Alembert:} \quad \sum_i \left(\mathbf{F}_i - m_i \frac{\mathrm{d}^2 \mathbf{r}_i}{\mathrm{d}t^2} \right) \delta \mathbf{r}_i = 0; \tag{1.33}$$

$$\text{in equilibrium:} \quad \sum_i \mathbf{F}_i \delta \mathbf{r}_i = 0; \tag{1.34}$$

$$\text{if } F_i \text{ conservative:} \quad \delta U = 0, \tag{1.35}$$

where U is the total potential energy.

The last equation $\delta V = 0$ summarises in only four symbols all the equilibrium questions of point mechanics. A machine may be arbitrarily complicated, with struts between the different masses, and rails on which the masses must move: nevertheless with this machine in equilibrium it is still true that a quite small displacement of any part cannot change the total potential U: the principle of virtual work. So this part of theoretical physics is seen to be not only elegant, but also practical. The law of levers is a particularly simple application: if the left-hand arm of a balance has length a and the righthand one length b, then the changes in height with a small rotation are as $a : b$. The potential energies $m_a g z$ and $m_b g z$ do not change in sum if $m_a g a = m_b g b$ or $m_a a = m_b b$. As an example for d'Alembert we can take Atwood's machine in Fig. 1.10: two point masses hang from a string which passes over a frictionless pulley. With what acceleration does the heavier mass sink?

Since the length of the string is constant, we have $\delta z_1 = -\delta z_2$ for the virtual displacements in the z-direction (upwards). The imposed gravity forces in the z-direction are $F_1 = -m_1 g$ and $F_2 = -m_2 g$. Hence we have

$$0 = \sum_i \left(F_i - m_i \mathrm{d}^2 z_i / \mathrm{d}t^2 \right) \delta z_i$$
$$= \left(-m_1 g - m_1 \mathrm{d}^2 z_1 / \mathrm{d}t^2 \right) \delta z_1 + \left(-m_2 g - m_2 \mathrm{d}^2 z_2 / \mathrm{d}t^2 \right) \delta z_2$$
$$= \delta z_1 \left(-m_1 g + m_1 a + m_2 g + m_2 a \right)$$

for arbitrary δz_1. So the contents of the brackets must be zero:

$$a = -g(m_2 - m_1)/(m_2 + m_1),$$

which as a clearly sensible result confirms the d'Alembert formalism.

In the next section we present this formalism in more detail; even this last section could be counted as analytical mechanics.

1.3 Analytical Mechanics

In this section we present the discussion, already begun, in more general formal methods. Later in quantum mechanics we shall become acquainted with their practical uses, e.g., the Hamilton function of position and momentum.

1.3.1 The Lagrange Function

(a) Generalised Coordinates and Velocities
Now we renumber the coordinates of all the N particles thus: instead of x_1, y_1, z_1, x_2, y_2, z_2, ..., x_N, y_N, z_N we write x_1, x_2, x_3, x_4, x_5, x_6, ..., x_{3N-1}, and x_{3N}. Now d'Alembert's principle from (1.33) has the form

$$\sum_i \left(F_i - m_i d^2 x_i / dt^2 \right) \delta x_i = 0.$$

These coordinates x_i, however, are not very convenient if constraints limit the motions. Then we should rather use generalised coordinates $q_1, q_2, ..., g_f$, if there are $3N - f$ restrictive conditions and hence f "degrees of freedom". These generalised coordinates should automatically fulfill the restrictive conditions, so that on inserting any numerical values for the q_μ there is no violation of the restrictive conditions, while on the other hand the declaration of all the q_μ completely specifies the system. If, for example, a motion follows a plane circular orbit with radius R, then instead of the traditional coordinates x_1 and x_2 with the condition $x_1^2 + x_2^2 = R^2$ it is much simpler to write the angle ϕ as the single generalised coordinate q. These generalised coordinates therefore do not necessarily have the dimension of a length; we usually restrict ourselves in practice to lengths and angles for the q_μ.

(b) Lagrange Equation of the Second Kind
The d'Alembert's principle mentioned above can now be rewritten in the new variables q_μ. For economy of effort we give the result immediately:

$$\frac{d\left[\partial L / \partial \dot{q}_\mu \right]}{dt} = \frac{\partial L}{\partial q_\mu}, \tag{1.36}$$

where the *Lagrange Function* $L = T - U$ is the difference between the kinetic and the potential energies, written as a function of the q_μ and their time derivatives. It is easy to decide where the dot for timewise differentiation occurs in (1.36) from dimensional considerations; and if one does not believe the whole equation it is easy to demonstrate it using the example $L = \Sigma_i m_i v_i^2/2 - U(x_1, x_2, ..., x_{3N})$, in the absence of restrictive conditions (hence $q_\mu = x_i$, and $v_i = dq_\mu/dt$). Then we obtain from (1.36): $m_i dv_i/dt = -\partial U/\partial x_i$, as required by Newton. If there are restrictive conditions, then they are elegantly eliminated from the Lagrange equation of the second kind (1788) by concealing them in the definition of the generalised coordinates $q\mu$.

Accordingly one proceeds in general as follows:

- choice of coordinates q_μ corresponding to the restrictive conditions;
- calculation of dx_i/dt as a function of q_μ and dq_μ/dt;
- substitution of the results in the kinetic energy T;
- calculation of the potential energy U as a function of the q_μ;
- derivation of $L = T - U$ with respect to q_μ and dq_μ/dt, substitution in (1.36).

We have therefore found a general method of calculating systems with arbitrarily complex restrictive conditions. In practice it often looks simpler than these general rules: for the pendulum of length l the coordinate q is the angle ϕ, the kinetic energy is $mv^2/2 = ml^2(d\phi/dt)^2/2$ and the potential energy is $-mgl \cos\phi$, if $\phi = 0$ is the rest position.

We accordingly have

$$L = \frac{1}{2}ml^2\dot{\phi}^2 + mgl \cos\phi,$$

so that (1.36) gives the usual pendulum equation $ml^2 d^2\phi/dt^2 = -mgl \sin\phi$ from Sect. 1.1.3b. Lagrange turns out to be correct.

(c) the Hamilton Principle of Least Action

We have here an extremal principle similar to many others in physics: the actual motion of a system is such that the action W is extremal, i.e. it is either a maximum or a minimum, when one considers all the possible motions from a specified starting point "1" to a specified endpoint "2". Here action is defined by the integral

$$W = \int_{t_1}^{t_2} L\left(q_\mu, \dot{q}_\mu\right) dt$$

along the motion path $q_\mu = q_\mu(t), \dot{q}_\mu = \dot{q}_\mu(t)$. With some calculation, and application of (1.36) and of partial integration one can show that with fixed endpoints "1" and "2":

$$\delta W = 0. \tag{1.37}$$

This Hamilton principle (1834) accordingly states that the action does not change if one alters the actual motion of the system very slightly. Vanishing of small variations is a well known characteristic of a maximum or a minimum. To experts in variational analysis (1.36) is readily recognised as the indication of an extremal principle.

Similarly, light "moves" in such a way that another integral, namely the traveling time, is minimal: Fermat's principle. From this follows, for example, the principle of geometric optics.

1.3.2 The Hamilton Function

It seems strange that the Lagrange function L is the difference and not the sum of the kinetic and the potential energies. This is different in the Hamilton function $H = T + U$; so this does not differ from the total energy, only we write it as a function of the (generalised) coordinates and momenta: $L = L(x, v)$, but $H = H(x, p)$ for a particle with position x, velocity v and momentum $p = mv$ in one dimension. The partial derivative d/dx accordingly leaves unchanged the velocity v in L, but the momentum p in H.

In case constraints are again present we define a generalised momentum

$$p_\mu = \partial L / \partial \dot{q}_\mu,$$

which in the absence of constraints coincides with the ordinary momentum $m dq_\mu/dt$. The Lagrange equation of the second kind now has the form $d p_\mu/dt = \partial L/\partial q_\mu$. Accordingly if a coordinate q_μ does not appear in the Lagrange function L of the system under consideration, so that L is invariant to changes in the variable q_μ, then the corresponding momentum p_μ is constant. For every invariance with respect to a continuous variable q_μ there accordingly is a conservation law. This was demonstrated more rigorously by Emmy Noether in 1918. Thus the constancy of the angular momentum follows from invariance with respect to a rotation of the total system, and the invariance of the total momentum from invariance with respect to a translation, as discussed in Sect. 1.2.1.

The total time derivative of the Lagrange function L is given by

$$\frac{dL}{dt} = \sum_\mu \left(\frac{\partial L}{\partial q_\mu} \frac{dq_\mu}{dt} + \frac{\partial L}{\partial \dot{q}_\mu} \frac{d\dot{q}_\mu}{dt} \right)$$

$$= \sum_\mu \left[\frac{d(\partial L/\partial \dot{q}_\mu)}{dt} \frac{dq_\mu}{dt} + \frac{\partial L}{d\dot{q}_\mu} \frac{d\dot{q}_\mu}{dt} \right]$$

$$= d\left(\sum_\mu \dot{q}_\mu \frac{\partial L}{\partial \dot{q}_\mu} \right) / dt.$$

Since the energy $E = -L + \Sigma_\mu \dot{q}_\mu \partial L/\partial \dot{q}_\mu$ we therefore have $dE/dt = 0$: the energy is constant. The fact that this E is actually the total energy $T + U$, shows that U is independent of the velocities, whereas T depends quadratically on the (generalised) velocities dq_μ/dt, and hence

$$\sum_\mu \dot{q}_\mu p_\mu = \sum_\mu \dot{q}_\mu \frac{\partial T}{\partial \dot{q}_\mu} = 2T.$$

We can therefore summarise as follows:

$$p_\mu = \frac{\partial L}{\partial \dot{q}_\mu}, \quad \dot{p}_\mu = \frac{\partial L}{\partial q_\mu}, \quad T + U = E = H = \sum_\mu p_\mu \dot{q}_\mu - L, \tag{1.38}$$

and this energy E is constant:

$$\frac{dE}{dt} = 0. \tag{1.39}$$

The energy is conserved here, because external forces and time dependent potentials were neglected.

Comparing now the differential $dH = \Sigma_\mu (\partial H/\partial q_\mu)dq_\mu + (\partial H/dp_\mu)dp_\mu$ with the analogous differential dL, and taking account of (1.38), we find the *canonical equations*

$$\dot{p}_\mu = -\frac{\partial H}{\partial q_\mu}, \quad \dot{q}_\mu = \frac{\partial H}{\partial p_\mu}, \quad H = H(q_\mu, p_\mu). \tag{1.40}$$

It is evident from the one-dimensional example of the free particle, $H = p^2/2m$, that these equations lead to the correct results $dp/dt = 0$, $dq/dt = p/m$. One also finds from this example where the minus sign is needed in (1.40).

As already mentioned, the Hamilton function plays an important role in quantum mechanics. The so-called commutator of quantum mechanics resembles the Poisson bracket of classical physics, defined by

$$\{F, G\} = \sum_\mu \left(\frac{\partial F}{\partial q_\mu} \frac{\partial G}{\partial p_\mu} - \frac{\partial F}{\partial p_\mu} \frac{\partial G}{\partial q_\mu} \right), \tag{1.41}$$

where F and G are any functions dependent on the positions q and the momenta p. Using the chain rule of differentiation it then follows that

$$\frac{dF}{dt} = \{F, H\} \tag{1.42}$$

just as the timewise variation of the quantum mechanical average value of a quantity F is given by the commutator $FH - HF$ (where F and H are "operators", i.e. a sort of matrices).

As example we take once again the one-dimensional harmonic oscillator: $T = mv^2/2$, $U = Kx^2/2$, with no restrictions, so that $q = x$, $p = mv$. Then the Hamilton function is

$$H(q, p) = p^2/2m + Kq^2/2.$$

From the canonical equations (1.42) it follows that $dp/dt = -Kq$ and $dq/dt = p/m$, which is correct. From (1.42), with $F = p$ in the Poisson bracket, it follows that

$$dp/dt = \{p, H\} = \frac{\partial p}{\partial q}\frac{\partial H}{\partial p} - \frac{\partial p}{\partial p}\frac{\partial H}{\partial q} = -\partial H/\partial q = -Kq.$$

which is also a correct result. We have thus successfully transcribed the simple law that force = mass times acceleration into a more complicated form, but one which is also more elegant, and suitable for the reader interested in practical applications in quantum mechanics. The next section, however, presents a different application.

1.3.3 Harmonic Approximation for Small Oscillations

A very commonly used approximation in theoretical physics is the harmonic approximation, where one develops a complicated function as a Taylor series and then truncates the series after the quadratic term. Applied to the potential energy U of a particle this gives

$$U(x) = U_0 + x\,dU/dx + (x^2/2)d^2U/dx^2 + ...,$$

where U_0 and dU/dx drop out if we take the origin of coordinates at the minimum of the energy $U(x)$. The Hamilton function is then $H = p^2/2m + Kx^2/2$ with $K = d^2U/dx^2 + ...$ (derivatives at the point $x = 0$), i.e. the well known function of the harmonic oscillator. In a solid body there are many atoms, which exert complicated forces upon each other. If we develop the total potential energy U about the equilibrium position of the atoms and truncate this Taylor series after the quadratic term, then this harmonic approximation leads to a large number of coupled harmonic oscillators. These are the lattice vibrations or phonons of the solid body. Before we mathematically decouple these 10^{24} oscillators, we must first learn with just two oscillators.

(a) Two Coupled Oscillators
Let two point masses of mass m be connected to one another by a spring, and connected to two rigid walls, each by a further spring (Fig. 1.11). The three springs all have the force constant K. Let the system be one-dimensional, the coordinates x_1 and x_2 giving the displacements of the two point masses from their rest positions. Then the Hamilton function, with $v_i = dx_i/dt$, is:

$$H = (m/2)\left[v_1^2 + v_2^2\right] + (K/2)\left[x_1^2 + x_2^2 + (x_1 - x_2)^2\right].$$

Fig. 1.11 Two coupled one-dimensional oscillators between two fixed walls. All three spring constants are equal

The kinetic energy is here a straight sum of two quadratic terms, but the potential energy on account of the coupling is proportional to $(x_1 - x_2)^2$. What is to be done about it?

Although there are no restrictive conditions here, we make use of the possibility discussed above of mathematically simplifying ("diagonalising") the problem by appropriate choice of coordinates q_μ. Thus, with $q_1 = x_1 + x_2$ and $q_2 = x_1 - x_2$, so that $x_1 = (q_1 + q_2)/2$ and $x_2 = (q_1 - q_2)/2$, we obtain

$$H = \frac{m}{4}\left[\dot{q}_1^2 + \dot{q}_2^2\right] + \frac{K}{4}\left[q_1^2 + 3q_2^2\right] = H_1^{osc} + H_2^{osc},$$

where H_1^{osc} depends only on q_1 and $\dot{q}_1$ and has the structure of the Hamilton function of a normal harmonic oscillator, similarly H_2. With the generalised momenta

$$p_i = \partial L/\partial \dot{q}_i = \partial H/\partial \dot{q}_i = m\dot{q}_i/2$$

and the canonical equations (1.40)

$$(m/2)\mathrm{d}^2 q_i/\mathrm{d}t^2 = \dot{p}_i = -\partial H/\partial q_i$$

we find the two equations of motion

$$m\mathrm{d}^2 q_i/\mathrm{d}t^2 = -K_i q_i$$

with $K_1 = K$ and $K_2 = 3K$. They are solved by $q_1 \sim \exp(\mathrm{i}\omega t)$ and $q_2 \sim \exp(\mathrm{i}\Omega t)$ with $\omega^2 = K/m$ and $\Omega^2 = 3K/m$. If $q_2 = 0$, so that $x_1 = x_2$, then the system oscillates with the angular velocity ω; if on the other hand $q_1 = 0$, so that $x_1 = -x_2$ then it oscillates with the larger $\Omega = \omega\sqrt{3}$. The masses therefore oscillate together with a lower frequency than if they swing against each other. In solid state physics one speaks of acoustic phonons when the vibrations are sympathetic, and of optical phonons when they are opposed. The general oscillation is a superposition (addition, or linear combination) of these two normal oscillations. The essential aspects of the harmonic vibrations in a solid body are therefore represented by this simple example; the next section does the same, only in a more complicated case.

(b) Normal Vibrations in the Solid State

We now calculate in a similar way the vibration frequencies of the atoms in a solid body which has atoms of mass m and only one type. Let the equilibrium position of the

i-th atom be r_i^0, and let $q_i = r_i - r_i^0$ be the displacement from equilibrium. We expand the potential U quadratically ("harmonic approximation") about the equilibrium position $q_i = 0$ and again number the coordinates i from 1 to $3N$:

$$U(q) = U(0) + \sum_{ik} \left(\partial^2 U / \partial q_i \partial q_k \right) q_i q_k / 2,$$

since the first derivatives vanish at equilibrium (minimum of the potential energy U). With the "Hesse matrix"

$$K_{ik} = \partial^2 U / \partial q_i \partial q_k = K_{ki}$$

the Hamilton function has the form

$$H = U(0) + \sum_i \left(p_i^2 / 2m \right) + \sum_{ik} K_{ik} q_i q_k / 2.$$

The canonical equation (1.40) then gives

$$-m \mathrm{d}^2 q_j / \mathrm{d}t^2 = -\dot{p}_j = \partial H / \partial q_j = \partial U / \partial q_j = \sum_k K_{jk} q_k,$$

which can of course also be derived directly from

$$\text{mass} \cdot \text{acceleration} = \text{force} = -\operatorname{grad} U.$$

(In the differentiation of the double sum there are two contributions, one from $i = j$ and the other from $k = j$; since $K_{ik} = K_{ki}$ the two terms are equal, so the factor 1/2 disappears.) For this system of linear differential equations (constant coefficients) we try the usual exponential solution: $q_j \sim \exp(i\omega t)$. This leads to

$$m\omega^2 q_j = \sum_k K_{jk} q_k. \tag{1.43}$$

Mathematicians recognise that on the right-hand side the $3N$-dimensional vector with the components $q_k, k = 1, 2, ..., 3N$ is multiplied by the Hesse matrix $\mathcal{K}$ of the K_{jk} and that the result (on the left-hand side) should be equal to this vector, to within a constant factor $m\omega^2$. Problems of this type

$$\text{factor} \cdot \text{vector} = \text{matrix} \cdot \text{vector}$$

are called eigenvalue equations (here the eigenvalue of the matrix is the factor, and the vector is the eigenvector). In general the equation system matrix $\cdot$ vector $= 0$ has a solution (with a vector which is not null) only if the determinant of the matrix is zero. If $\mathcal{E}$ is the unit matrix of Kronecker deltas, so that $\mathcal{E}_{jk} = \delta_{jk} = 1$ for $j = k$ and $= 0$ otherwise, then eigenvalue equations have the form

$$\text{(matrix } - \text{ factor} \cdot \mathcal{E}) \cdot \text{ vector } = 0,$$

which leads to

$$\text{determinant of (matrix } - \text{ factor} \cdot \mathcal{E}) = 0,$$

as the condition for a solution. The determinant det of a two-dimensional matrix is

$$\det \begin{pmatrix} a & b \\ c & d \end{pmatrix} = ad - bc;$$

the reader will find larger matrices treated in books on linear algebra.

In the case of rigid body vibrations we therefore have to set to zero the determinant of a $3N$-dimensional matrix:

$$\det(K - m\omega^2 \mathcal{E}) = 0. \tag{1.44}$$

From linear algebra it is well known that the eigenvalues of a symmetric matrix ($K_{jk} = K_{kj}$) are real and not complex. If the potential energy in equilibrium is a minimum, which it must be for a stable equilibrium, then no eigenvalues $m\omega^2$ are negative, so that ω is also not imaginary. We therefore have true vibrations, and not disturbances decaying exponentially with time.

The so-called secular equation (1.44) is a polynomial of degree $3N$, which is really troublesome to calculate with $N = 10^{24}$ atoms. It is easier if one assumes that in equilibrium all atoms lie in positions on a periodic lattice. Then one makes the trial solution of a plane wave:

$$q_j \sim \exp(i\omega t - i\boldsymbol{Q}\boldsymbol{r}_j^0), \tag{1.45}$$

where q_j is now a three-dimensional vector, $j = 1, 2, ..., N$, and $\boldsymbol{Q}$ indicates a wave vector. With this simplification the eigenvalue problem is reduced to that of a three-dimensional "polarisation vector" $\boldsymbol{q}$ with the associated eigenvalue $m\omega^2$, both of which depend on the wave vector $\boldsymbol{Q}$. (See textbooks on solid state physics.) To determine the eigenvalues of a 3×3 matrix leads to an equation of the third degree; in two dimensions one has a quadratic equation to solve. Typical solutions for the frequency ω as a function of the wave vector $\boldsymbol{Q}$ in three dimensions have the form of Fig. 1.12, where A stands for "acoustic" (sympathetic vibrations), O for "optical" (opposed vibrations), and L for longitudinal (displacement $\boldsymbol{q}$ in the direction of the wave vector $\boldsymbol{Q}$) and T for transverse. With only one sort of atom there are only three acoustic branches (left), with two different sorts of atoms there are also three optical branches (right). In quantum mechanics these vibrations are called phonons.

(c) Linear Chains

We now wish to calculate explicitly the frequency spectrum $\omega(Q)$ in one dimension, i.e. in an infinitely long chain of identical point masses m. Between the points j and

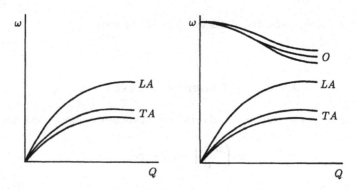

Fig. 1.12 Typical phonon spectra in three-dimensional crystals

$j + 1$ there is a spring with the force constant K; if neighbouring point masses are separated by the distance a the spring force is zero and the atoms are in equilibrium: $x_j^0 = aj$ for $-\infty < j < +\infty$.

The Hamilton function or total energy is then

$$H = \sum_j (p_j^2/2m) + \frac{K}{2} \sum_j (q_{j+1} - q_j)^2 = \sum_j (p_j^2/2m) + \sum_{jk} K_{jk} q_j q_k r/2$$

with $q_j = x_j - x_j^0$ and the matrix elements $K_{jk} = 0, -K, 2K, -K$ and 0 for $k < j - 1, k = j - 1, k = j, k = j + 1$, and $k > j + 1$, respectively. The trial solution of a plane wave (1.45) with wave vector Q, $q_j \sim \exp(i\omega t - iQaj)$, using (1.43), gives

$$m\omega^2 \exp(i\omega t - iQaj) = \sum_k K_{jk} \exp(i\omega t - iQak) \text{ or}$$

$$m\omega^2 = \sum_k K_{jk} \exp(iQa(j - k))$$

$$= -K \exp(iQa) + 2K - K \exp(-iQa)$$

$$= -K(\exp(iQa/2) - \exp(-iQa/2))^2 = 4K \sin^2(Qa/2)$$

so that

$$\omega = \pm 2(K/m)^{1/2} \sin(Qa/2). \tag{1.46}$$

To be meaningful the wave vector Q is limited to the region $0 \le |Q| \le \pi/a$, because in a periodic chain the wave vectors Q and $Q + 2\pi/a$, for example, are completely equivalent (between the atoms there is nothing that could move). In this so-called Brillouin zone between $Qa = 0$ and $Qa = \pi$ the sine in (1.46) rises from 0 to 1, just as it does schematically for the longitudinal acoustic phonon branch in Fig. 1.12.

1.4 Mechanics of Rigid Bodies

The theme of this section is the motion of solid bodies as entities. With an iron plate we no longer consider this plate as a point mass, as in Sects. 1.1 and 1.2, nor as a system of 10^{25} or more inter-vibrating atoms, as on the previous pages, but we ask, for example, what forces act on the plate if it is attached to a motor and then rotated. Why do gyroscopes behave in the way they do? In general, then, we consider rigid bodies, in which the distances and the angles between different atoms are *fixed* (more precisely: in which the changes in distances and angles are negligible).

1.4.1 Kinematics and Inertia Tensor

(a) Rotations

If a rigid body rotates about an axis with the angular velocity $\omega = \partial\phi/\partial t$, then the vector ω lies in the direction of the axis (Fig. 1.5). The body rotates in the clockwise direction when regarded in the direction of $+\omega$: the rule of the thumb of the right hand. The fact that here right has precedence over left is due, not to politics, but to the cheating of physicists: they regard certain asymmetric 3×3 matrices as axial vectors, although they are not true vectors. These pseudo-vectors correspond to cross-products, magnetic fields, and vectors, such as ω itself, defined by the direction of rotation. With the definition of tensors later on, and in Sect. 2.3.2 (Relativistic Electrodynamics), we shall see more of these imposters.

The velocity v of a point on the rotating rigid body at a position r relative to the origin is the cross-product

$$v = \omega \times r, \tag{1.47}$$

assuming (as we shall always assume in future) that the origin of coordinates lies on the axis of rotation. Not only v but also r are genuine polar vectors, ω and the cross-product of two polar vectors are axial vectors. Axial vectors, unlike polar vectors, change their sign if the x-, y- and z-axes all change their signs ("inversion" of the coordinate system). The two sign changes in ω and the cross-product therefore cancel out in (1.47). In general (1.47) can best be made clear by considering points on a plane which is at right-angles to the axis of rotation; points on the rotation axis have no velocity v.

If one holds the axle of the front wheel of a bicycle, sets it spinning rapidly, and then tries to turn the axle into a different direction, one will notice the tendency of the axle to turn at right-angles to the force being exerted on it. This evasion at right-angles is easy to explain in principle: the timewise variation of the angular momentum L is according to (1.13) equal to the torque M. This again is $r \times f$; if then the force f is applied perpendicular to the axle at the position r, then the torque M and the change in the angular momentum are perpendicular both to the axle and to the force (Fig. 1.13). We should find this easy to understand. In the following section

Fig. 1.13 Simple explanation of the perpendicular evasion of the applied force F. The angular momentum L points upwards. On the *left* the gyro is viewed from above, on the *right* from the side. L changes in the direction of the torque M

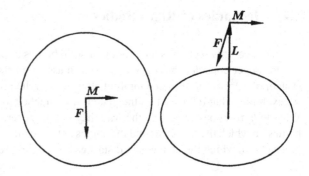

we shall replace this qualitative explanation by a more precise, but unfortunately more complicated, argument.

The gyrocompass is a practical application. Since the earth is not an inertial system, but spins daily on its axis, this terrestrial rotation exerts a torque on every rotating rigid body having its axis of rotation fixed to the earth's surface. If instead the axis of rotation is suspended in such a way that it can rotate horizontally to the earth's surface, but not vertically, then the torque from the terrestrial rotation leads in general to the above mentioned deflection perpendicular to the axis of the gyroscope. This continuing deflection of the axis of the gyroscope ("precession") causes frictional losses; gradually the gyroscope axis sets itself in the north-south direction, where the precession no longer occurs. The flight of the boomerang is also based on the gyroscopic effect; its demonstration by a theoretical physicist in a fully occupied lecture hall, however, has certain disadvantages.

(b) Angular Momentum and Inertia Tensor

For the cross-product with a cross-product we have the transformation $a \times (b \times c) = b(ac) - c(ab)$ into scalar products. We apply this rule to the angular momentum L_i, of the ith atom or mass element:

$$L_i m_i = r_i \times v_i = r_i \times (\omega \times r_i)$$

$$= \omega(r_i r_i) - r_i(r_i \omega) = \omega r_i^2 - \sum_{\nu=1}^{3} \emptyset \omega_\nu r_{i\nu} r_i,$$

or in components ($\mu, \nu = 1, 2, 3$):

$$L_{i\mu}/m_i = \omega_\mu r_i^2 - \sum_\nu \omega_\nu r_{i\mu} r_{i\nu} = \sum_\nu \omega_\nu \left(r_i^2 \delta_{\mu\nu} - r_{i\mu} r_{i\nu}\right)$$

with again the Kronecker delta $\delta_{\mu\nu} = 1$ for $\mu = \nu$ and $= 0$ otherwise. For the components of the total angular momentum $L = \Sigma_i L_i$ we have

$$L_\mu = \sum_\nu \omega_\nu \Theta_{\mu\nu} \text{ or } L = \Theta\omega, \quad \Theta_{\mu\nu} = \sum_i m_i \left(r_i^2 \delta_{\mu\nu} - r_{i\mu} r_{i\nu}\right). \qquad (1.48)$$

The matrix Θ of the $\Theta_{\mu\nu}$ so defined is called the *inertia tensor*. Overlooking this matrix property the relation $L = \Theta\omega$ for the rotation of a rigid body is quite analogous to the momentum definition $p = mv$ for its translational motion. Tensors are "true" matrices with physical meaning. More precisely: a vector for a *computer program* is any combination of (in three dimensions) three numbers, e.g., the number triplet consisting of, in the first place the Dow Jones Index from Wall Street, in the second place the body weight of the reader, and in the third place the size of the university budget. For *physics* this is gibberish, whereas, for example, the position vector is a true vector. For the physicist, moreover, true vectors are those number triplets which transform like a position vector under a rotation of the coordinate system. Similarly, not every square array of numbers which a computer could store as a matrix would be regarded by a physicist as a tensor. Tensors are only those matrices whose components are transformed under a rotation of the coordinate system in such a way that the tensor before and after links the same vectors. Accordingly, for true vectors and tensors the relation: vector$_1$ = tensor · vector$_2$ is independent of the direction of the coordinate axes. Only then do equations such as (1.48) make sense.

Since the inertia tensor Θ is symmetric, $\Theta_{\mu\nu} = \Theta_{\nu\mu}$, it has only real eigenvalues. Moreover, for all symmetric tensors one can choose the eigenvectors so that they are mutually perpendicular. If we therefore set our coordinate axes in the directions of these three eigenvectors, then any vector lying in the x-axis will, after multiplication by the tensor Θ, again lie in the x-axis, but with its length multiplied by the first eigenvalue, called Θ_1. Similarly, any vector in the y-direction, after application of the matrix Θ, will be stretched or shortened by the factor Θ_2, without change of direction. The third eigenvalue Θ_3 applies to vectors in the z-direction. General vectors are made up of their components in the x-, y- and z-directions, and after multiplication by Θ are again the sum of their three components multiplied by Θ_μ. Accordingly in the new coordinate system with its axes in the direction of the eigenvectors we have

$$L = \begin{pmatrix} \Theta_1 & 0 & 0 \\ 0 & \Theta_2 & 0 \\ 0 & 0 & \Theta_3 \end{pmatrix} \cdot \omega = \Theta\omega \quad \text{or} \quad L_\mu = \Theta_\mu \omega_\mu \qquad (1.49)$$

for $\mu = 1, 2, 3$. The tensor Θ therefore has a diagonal form in the new coordinate system; outside the diagonal the matrix consists of zeros.

Mathematicians call this choice of coordinate system, possible with any symmetric matrix, its *principal axes* form; one has referred the tensor to its principal axes, or "diagonalised" it. Physicists call the eigenvalues Θ_1 of the inertia tensor the *principal moments of inertia*.

If one uses these principal axes one has

$$\Theta_\mu = \Theta_{\mu\mu} = \sum_i m_i \rho_i^2 \quad \text{with} \quad \rho_i^2 = r_i^2 - r_{i\mu}^2, \qquad (1.50)$$

where ρ is the distance of the position r from the μ-axis; when $\mu = 1$, i.e. the x-axis, we accordingly have $\rho^2 = x^2 + y^2 + z^2 - x^2 = y^2 + z^2$, as it should be. If one rotates the rigid body about a fixed axle, not just about an imaginary axis, then (1.50) is likewise valid with the axle in place of the μ-axis: $L = \Sigma_i m_i \rho_i^2 \omega$. In this case one calls $\Sigma_i m_i \rho_i^2$ the moment of inertia ϑ; ϑ is then a number and no longer a tensor. Do you still remember the Steiner rule? If not, have you at least come across Frisbee disks?

(c) Kinetic Energy

If the centre of mass of the rigid body of mass M does not lie at the origin then its kinetic energy is $T' = T + P^2/2M$, where P is the total momentum and T the kinetic energy in the coordinate system whose origin coincides with the centre of mass of the rigid body. It is therefore practical, here and elsewhere, to use the latter system at once and calculate T.

We have

$$2T = \sum_i m_i v_i^2 = \sum_i m_i v_i (\omega \times r_i) = \omega L,$$

where at the end we have again applied the "triple product" formula (volume of a parallelepiped) $a(b \times c) = b(c \times a) = c(a \times b)$. Hence:

$$2T = \omega L = \omega \Theta \omega = \sum_{\mu\nu} \omega_\mu \Theta_{\mu\nu} \omega_\nu = \omega_1^2 \Theta_1 + \omega_2^2 \Theta_2 + \omega_3^2 \Theta_3, \tag{1.51}$$

where the last relation is valid only in the principal axes system of the body. If the body rotates with moment of inertia ϑ about a fixed axis, (1.51) is simplified to $2T = \vartheta \omega^2$. Since in the absence of external forces the kinetic energy is constant, $\Sigma_{\mu\nu} \omega_\mu \Theta_{\mu\nu} \omega_\nu$ is therefore constant. This condition describes an *ellipsoid of inertia* in ω-space. If the three principal moments of inertia are equal this "ellipsoid" clearly degenerates into a sphere. One usually calls any rigid body with three equal principal moments of inertia a "spherical gyroscope", although besides the sphere a homogeneous cube also qualifies for this. "Symmetric" *gyroscopes* are those with two of the three principal moments of inertia Θ_μ equal.

The angular momentum L is according to (1.49) the gradient (in ω-space) of the kinetic energy T according to (1.51) and is therefore normal to this ellipsoid of inertia; in general grad f is normal to the surface defined by $f(r) = $ const. In general, therefore, as shown by Fig. 1.14, the vectors ω and L are not parallel. Only when the ellipsoid degenerates into a sphere, i.e. all three moments of inertia are equal, are ω and L always parallel. Mathematicians will see this directly from the relation (1.49).

1.4.2   Equations of Motion

(a) Fundamentals

The rigid body is in equilibrium only if no external torque nor any external force acts on it. This is not quite trivial for individual point masses, as is a well known

Fig. 1.14 Two-dimensional illustration of (1.49) and (1.51). If the inertia ellipse (in principal axis form $\Theta_1\omega_1^2 + \Theta_2\omega_2^2 = 2T$) does not degenerate into a circle, the vectors ω and L are in general not parallel

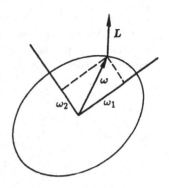

fact from daily experience, since enormous constraining forces, and perhaps torques also, act between the atoms of the rigid body. However, these all balance out, as one sees from the principle of virtual work (1.32). If the whole body is displaced through the distance δR and rotated through the angle $\delta\phi$, then $\delta r_i = \delta R + \delta\phi \times r_i$, so since these virtual displacements do no work we have:

$$0 = \sum_i F_i \delta r_i = \delta R \sum_i F_i + \delta\phi \sum_i r_i \times F_i$$

for all small δR and $\delta\phi$. Then the sums must also vanish: $\Sigma_i F_i = 0 = \Sigma_i r_i \times F_i$. Accordingly the total force and also the total torque vanish.

If an external force F and an external torque M act on the rigid body, these determine the changes in the total momentum P and the total angular momentum L, precisely because the inner forces and torques all cancel out:

$$F = \frac{dP}{dt}, \quad M = \frac{dL}{dt} = \frac{d(\Theta\omega)}{dt} \tag{1.52}$$

in an inertial system. These are six equations for six unknowns, so we find ourselves in a promising situation.

(b) Euler's Equations

If a body rotates, then all its principal axes rotate with it, and also the entire inertia tensor. We consider this body from an inertial system under the influence of an external torque M and denote by e_μ the unit vectors in the directions of the principal axes (eigenvectors of the inertia tensor). Then these e_μ change in time at the rate $de_\mu/dt = \omega \times e_\mu$, $\mu = 1, 2, 3$. The angular momentum is then seen from the inertial system, taking account of the diagonal form (1.49) of the tensor Θ, to be:

$$L = \Theta\omega = \sum_\mu \Theta_\mu \omega_\mu e_\mu.$$

Here we use $\omega = \Sigma_\mu \omega_\mu e_\mu$, which sounds trivial but establishes that now the three ω_μ are the components relative to the e_μ system of reference fixed in the body, and not relative to the inertial system.

For the time derivative of L we therefore have

$$M = \dot{L} = \sum_\mu \left(\Theta_\mu \dot{\omega}_\mu e_\mu + \Theta_\mu \omega_\mu \dot{e}_\mu \right).$$

Substituting

$$de_1/dt = (\omega_1 e_1 + \omega_2 e_2 + \omega_3 e_3) \times e_1 = \omega_3 e_2 - \omega_2 e_3,$$

and similar relations for the two other components, finally reduces the above expression for M to the Euler equations:

$$\begin{aligned}
M_1 &= \Theta_1 d\omega_1/dt + (\Theta_3 - \Theta_2)\,\omega_2\omega_3 \\
M_2 &= \Theta_2 d\omega_2/dt + (\Theta_1 - \Theta_3)\,\omega_3\omega_1 \\
M_3 &= \Theta_3 d\omega_3/dt + (\Theta_2 - \Theta_1)\,\omega_1\omega_2.
\end{aligned} \tag{1.53}$$

If one knows one of these equations, the others follow from it naturally by cyclic exchange of the indices: 1 by 2, 2 by 3, 3 by 1. In a spherical gyroscope all three Θ_μ are equal and hence we simply have

$$M = \Theta_\mu d\omega/dt.$$

A remarkable thing about these equations is first of all that they are not linear, but quadratic in ω. Since we often can only solve linear differential equations exactly, we program their simulation by the method already described in Sect. 1.1, for the case $M = 0$ (see Program EULER).

PROGRAM EULER
```
 10 input "omega="; w1, w2, w3
 20 dt = 0.01
 30 t1 =10.0
 40 t2 = 1.0
 50 t3 = 0.1
 60 d1 =dt*(t2-t3)/t1
 70 d2 =dt*(t3-t1)/t2
 80 d3 =dt*(t1-t2)/t3
 90 w1 =w1+d1*w2*w3
100 w2=w2+d2*w3*w1
110 w3=w3+d3*w1*w2
120 print w1, w2
130 goto 90
140 end
```

Whether linear or nonlinear, it is all the same to the BASIC program; it is more important that we must denote ω by w. The three principal moments of inertia are 10, 1 and 1/10, the time-step dt is 1/100. If we allow the body to rotate about a principal axis, e.g. by the input 0, 1, 0, then nothing changes at all. If we introduce small disturbances, however, inputing one of the three ω_μ as 1, the other two as 0.01, then the picture changes. A rotation about the principal axis with the greatest moment of inertia (here axis 1) is stable, i.e. the small disturbances in the two other components oscillate about zero and remain small, while ω_1 remains in the neighbourhood of 1. This is obtained by inputing 1.0, 0.01, 0.01. With the input 0.01, 1, 0.01, on the other hand, i.e. with a rotation about the axis with the middle moment of inertia, the body totters about all over the place: the initially small ω_3 becomes up to ten times larger, but above all ω_2 changes its sign. The initially dominant rotation component ω_2 is thus influenced quite critically by the small disturbances. Of course, none of these disturbances grows exponentially without limit (contrary to linear differential equations), since the kinetic energy must be conserved. The rotation about the third principal axis with the smallest moment of inertia is again stable.

One can try experimentally to demonstrate the stability (instability) for rotation about the axis with the greatest (middle) moment of inertia by skilful throws of filled matchboxes. But here one comes close to the border between accurate observation and hopeful faith. Instead, one can treat the Euler equations in harmonic approximation and theoretically distinguish clearly between instability (exponential increase of disturbances) and stability; Euler (1707–1783) knew no BASIC.

(c) Nutation

The stable rocking of a gyroscope without external torque is called *nutation*; we call *precession* the gradual rotation of the rotation axis under the influence of a weak torque. Other definitions also occur in the literature. We now calculate the nutation frequency, which we could observe empirically in the quantity ω_2 with the above computer program. We consider the symmetric gyroscope, $\Theta_1 = \Theta_2$, and assume that the gyroscope spins fast about the third axis with moment Θ_3, but that ω_1 and ω_2 are not exactly zero. In this stable case how do the two components ω_1 and ω_2 oscillate, i.e. how does the instantaneous axis of rotation rock, when seen from the rigid body? The ω_μ in (1.53) and here are still always the components in the principal axes system fixed in the rigid body.

The Euler equations with the abbreviation $\tau = (\Theta_1 - \Theta_3)/\Theta_1$ now become

$$d\omega_1/dt = \tau\omega_2\omega_3, \quad d\omega_2/dt = -\tau\omega_3\omega_1, \quad d\omega_3/dt = 0.$$

The principal component ω_3 therefore remains constant and we have

$$d^2\omega_1/dt^2 = \tau\omega_3 d\omega_2/dt = -r_2\omega_3^2\omega_1.$$

This is once again the equation of the harmonic oscillator and is similarly valid for ω_2. The well known solution is

$$\omega_\mu \sim e^{i\Omega t} \ (\mu = 1, 2), \quad \Omega/\omega_3 = \tau = (\Theta_1 - \Theta_3)/\Theta_1. \tag{1.54}$$

Accordingly, if the axis of rotation in the direction of ω does not exactly coincide with the body axis e_3 of the symmetrical gyroscope, so that ω_1 and ω_2 are not zero, then the axis of rotation wobbles with the nutation frequency Ω about the body axis e_3. This nutation frequency is proportional to the actual rotation frequency ω_3; the factor of proportionality is a ratio of the moments of inertia.

Since the three ω-components are measured from the rotating reference system fixed in the body, one can easily become dizzy. It is safer if we take the planet earth as the example of a rigid body. It is known that the earth is not a sphere, but is slightly flattened; the principal moment of inertia Θ_3 relating to the north-south axis is therefore somewhat greater than the other two in the equatorial plane. The reference system fixed in the body is now our longitude and latitude, familiar from maps. If we define the south pole by the direction of the principal moment of inertia e_3, the instantaneous axis of rotation will not coincide exactly with this pole, but will nutate about it. Since $\tau = 1/300$ the nutation frequency Ω must correspond to a period of about 300 days. Actually the pole is observed to wobble with a period of 427 days, as the earth is not a rigid body. Volcanic eruptions show that it is fluid inside.

If one observes the nutating symmetric gyroscope from the inertial system instead of from the system fixed in the body, it is no longer the axis e_3 of the body that is always in the same direction, but the angular momentum (provided that there is no external torque). Around this fixed direction of the angular momentum the axis e_3 of the body describes the "nutation cone". The instantaneous axis of rotation ω rotates on the "rest cone" or "herpolhode cone", while ω itself spins about the body axis on the "rolling cone" or "polhode cone".

(d) Precession

What happens when an external torque acts on a symmetric gyroscope? For example, this is the situation for a toy spinning-top whose tip rests on the ground and which is not perfectly upright. For the reader whose youth was so occupied with video-games that he had no time for such toys, Fig. 1.15 shows a sketch of this experimental apparatus.

If m is the mass of the top, R the distance of its centre of mass from its tip (point of support) and g the downward acceleration of terrestrial gravity, then the weight mg exerts the torque $m\mathbf{R} \times \mathbf{g}$ on the top. The vector $\mathbf{R}$ lies in the direction of the body axis (at least when the top is perfectly round), and this is in the direction of the angular momentum $\mathbf{L}$ if we neglect the nutation. (We thus assume that the top is spinning exactly about its axis of symmetry.) We therefore have

$$\frac{d\mathbf{L}}{dt} = \mathbf{M} = \omega_L \times \mathbf{L}, \tag{1.55}$$

where the vector ω_L, acts upwards and has the magnitude mRg/L. The solution of this equation is simple: the horizontal component of the angular momentum (and therefore also that of the body axis) rotates with the angular velocity ω_L about the

Fig. 1.15 Example of a
symmetric spinning-top in
the field of gravity. The
angular momentum L is
almost parallel to ω, the
torque M is perpendicular to
the plane of the paper

vertical. This slow rotation proportional to the external torque is called precession
(other names also occur in the literature for our nutation and precession). The magnitude of L and its vertical component L_3 accordingly remains constant. With real
tops, of course, there are also frictional forces. This explains why the top does not
fall over, but moves perpendicular to the direction in which it would be expected to
fall.

Another example of the application of (1.55) is the "Larmor precession" of magnetic moments ("spins"). In the classical atom an electron orbits around the nucleus,
and because of its electrical charge produces a circular electrical current and hence
a magnetic dipole moment μ. An atom therefore usually has a moment of inertia
and an angular momentum L, and also a magnetic dipole moment μ proportional
to L. (We shall learn later in electrodynamics that charges at rest cause electric
fields, while moving charges cause magnetic fields and oscillating charges waves.)
In a magnetic field B a magnetic dipole moment experiences a torque $B \times \mu$. Then
(1.55) is again valid, with the Larmor frequency $\omega_L = |B \times \mu|/L$. In the absence
of quantum mechanics, elementary magnets would therefore continually precess if
they were not exactly parallel to the magnetic field. The "gyromagnetic" ratio μ/L
is proportional to the ratio of the elementary electric charge e to the product of mass
m and the velocity of light $c : \omega_L = eB/mc$ in appropriate units.

Such effects are applied in spin resonance (NMR, MRI: since 1946) to the study
of solid bodies and biological macromolecules, but more recently also in medicine
to diagnosis without an operation and without X-radiography (NMR tomography,
NMR = Nuclear Magnetic Resonance).

Precession is also important for *horoscopes*. Because of the flattening of the earth
the gravitation of the sun exerts a torque on the earth, and the angular momentum of
the earth processes with a period of 26,000 years. Accordingly the agreement between
the stellar constellations and the calendar months becomes worse and worse with the
passage of time; every 26,000/12 years the signs of the zodiac move along by one
sign. Since the signs of the zodiac had already been fixed a long time ago, they are no
longer correctly placed today. Modern foretellers of the future therefore always have

Fig. 1.16 Motion of the peak of a *spinning-top* with weak (*bottom*) and with strong (*top*) precession, as well as steady nutation. This garland may be regarded as a laurel wreath (Nobel Prize substitute) fashioned for the reader

to read between the signs, casting the horoscope according to the average value of the two predictions of two neighbouring signs of the zodiac. In this way we arrived at the prediction that this textbook would be a great success.

Whether it concerns mechanical gyroscope or magnetic spin, the torque precesses with the angular velocity ω_L on a cone about the vertical, if the gravitational force or magnetic field acts downwards. This simple result holds only, of course, when both friction and nutation are neglected. If there is a weak nutation, since the symmetric gyroscope does not rotate exactly around the body axis e_3, the vector e_3 no longer moves on the cone, so its end no longer moves on a circle. Instead, the end of e_3 moves in a series of loops (strong amplitude of nutation) or waves (weak nutation) formed by the superposition of two circular motions (Fig. 1.16).

1.5   Continuum Mechanics

1.5.1   *Basic Concepts*

(a) Continua

Elastic solids, flowing liquids and drifting gases are the continua of this Section on *elasticity* and *hydrodynamics*. If in this sense a solid is not rigid, then one has actually to treat all the molecules separately. In a glass of beer there are about 10^{25} particles, and there are more congenial methods to go about this than to solve Newton's equations of motion for all of them simultaneously. Instead, we once again use an approximation: we average over many atoms. If we wish to describe the flow of air round a motor-car or the deformation of an iron plate supporting a heavy load, then in these engineering applications we are scarcely interested in the thermal motion of the air molecules or the high frequency phonons in the iron. We wish to find a mean velocity of the air molecules and a mean displacement of the atoms of iron from their equilibrium positions. We need therefore to average over a "mesoscopic" region, containing many molecules, but small compared with the deformation of the solid or with the distances over which the velocity of the fluid flow changes significantly.

Actually we do not really carry out this averaging; only in recent years has hydrodynamics been studied on the computer by the simulation of every individual atom. We accordingly restrict ourselves here to postulating that there is a mean deformation and a mean velocity. On this assumption we construct the whole of continuum mechanics, without actually calculating these mean values from the individual molecules. We shall later use similar tricks with Maxwell's equations in matter and in thermodynamics. If we do not know a quantity which would in principle be calculated from the individual molecules, then we give this quantity a name ("density", "viscosity", "susceptibility", "specific heat") and assume that it can be measured concurrently by experimental physics. We then work with this measured value, in order to predict other measured values and phenomena. This may be regarded as cheating, but this method has been vindicated over hundreds of years. A theory is generally called "phenomenological" if certain material properties are not calculated, but only measured experimentally.

Almost all the formulae in this section hold in common for gases, liquids and solids. In any case, one cannot always distinguish clearly between these phases, since iron plates can be deformed even more easily than glass, and at the critical point (see van der Waals' equation) the difference between vapour and liquid disappears. Nevertheless, when the discussion is about strain, the reader can think of a single-crystal solid; in velocity fields it is best to think of the flow of "incompressible" water, and shock waves can be envisaged in "compressible" air.

(b) Strain Tensor ε

In an elastic solid let u be the mean displacement of the molecules from the equilibrium configuration; u depends on the position r in the solid under consideration. (For liquids and gases u is the displacement from the position at time $t = 0$.) For sufficiently small distances r between two points in the solid we have the Taylor expansion:

$$u(r) = u(0) + \sum_k x_k \partial u / \partial x_k, \quad k = 1, 2, 3.$$

We define

$$\text{div } u = \frac{\partial u_1}{\partial x_1} + \frac{\partial u_2}{\partial x_2} + \frac{\partial u_3}{\partial x_3} \tag{1.56}$$

$$\text{curl } u = \left(\frac{\partial u_3}{\partial x_2} - \frac{\partial u_2}{\partial x_3}, \frac{\partial u_1}{\partial x_3} - \frac{\partial u_3}{\partial x_1}, \frac{\partial u_2}{\partial x_1} - \frac{\partial u_1}{\partial x_2} \right) \tag{1.57}$$

as the *divergence* and the *curl* of the quantity $u(r)$. Many authors write div u as the scalar product of the nabla operator $\nabla = (\partial/\partial x_1, \partial/\partial x_2, \partial/\partial x_3)$ with the vector u; in this sense curl u is the cross-product $\nabla \times u$. Many rules concerning scalar and cross-products are also valid here. The point of prime importance is that the curl is a vector, and the divergence is not.

After some manipulation the above Taylor expansion becomes

$$u(r) = u(0) + \text{curl}(u) \times r/2 + \varepsilon r \qquad (1.58)$$

with the *strain* tensor ε, a 3×3 matrix, defined by

$$\varepsilon_{ik} = (\partial u_i/\partial x_k + \partial u_k/\partial x_i)/2 = \varepsilon_{ki}. \qquad (1.59)$$

This shows clearly that the displacement u can be represented in small regions (r not too large) as a superposition of a translation $u(0)$, a rotation through the angle $\text{curl}(u)/2$, and a distortion or strain of the elastic solid. For the rigid solids of the previous section the distortion is absent, and $\text{curl}(u)$ is uniform over space.

Since the strain tensor ε is always symmetric, there is a rectangular coordinate system in which the matrix of the ε_{ik} is diagonal: $\varepsilon_{ik} = 0$ except when $i = k$. In this coordinate system the volume change ΔV of a distorted prism of length x, breadth y and height z is especially convenient to calculate, since now $\Delta x = \varepsilon_{11} x$, etc.:

$$\Delta V/V = [(x + \Delta x)(y + \Delta y)(z + \Delta z) - xyz]/xyz \approx \varepsilon_{11} + \varepsilon_{22} + \varepsilon_{33} = \text{Tr}(\varepsilon)$$

with the trace $\text{Tr}(\varepsilon) = \Sigma_i \varepsilon_{ii}$. Mathematicians have proved that the trace of a matrix does not change with a rotation of the coordinate system. The trace of the unit tensor $\mathcal{E}$, defined as the matrix of the Kronecker delta δ_{ik}, is trivially equal to 3. With the definition

$$\varepsilon = \varepsilon' + \text{Tr}(\varepsilon)\mathcal{E}/3$$

the strain tensor is partitioned into a shear ε' without volume change (since $\text{Tr}(\varepsilon') = 0$) and a volume change without change of shape (since it is proportional to the unit matrix). This analysis of the general displacement u into a translation, a rotation, a change of shape and a change of volume is very plausible even without mathematics.

(c) Velocity Field
In gases and liquids the displacement field $u(r)$ can be described as the displacement of the molecules from their positions at time $t = 0$; there is no equilibrium position. It is more appropriate, however, to talk of a mean velocity $v(r)$ of the molecules: $v = du/dt$. The velocity field v depends on the time t, as well as on the position r.

A clear distinction must be made between the total time derivative d/dt and the partial time derivative $\partial/\partial t$. This distinction can be clarified physically by considering the temperature T in a stream of water. If one measures it at a fixed position, e.g., at a bridge, then the position r is held constant and the measured rate of temperature change is consequently $\partial T/\partial t$. If, on the other hand, one drops the thermometer into the stream, so that it drifts along with the current, then one measures the heating or cooling of the portion of water in which the thermometer remains all the time it is drifting. This rate of change of temperature, with varying position, is therefore dT/dt.

Mathematically the two derivatives are connected via the temperature gradient grad T:

$$dT/dt = \partial T/\partial t + (\partial T/\partial x)(\partial x/\partial t) + (\partial T/\partial y)(\partial y/\partial t) + (\partial T/\partial z)(\partial z/\partial t)$$
$$= \partial T/\partial t + \sum_i v_i \partial T/\partial x_i = \partial T/\partial t + (v\,\mathrm{grad})T,$$

where $(v\,\mathrm{grad})$ is the scalar product of the velocity with the nabla operator ∇. Another notation for this operator $(v\,\mathrm{grad})$ is $(v \cdot \nabla)$; anybody who finds this operator notation difficult can always replace the expression $(v\,\mathrm{grad})T$ by $\Sigma_i v_i \partial T/\partial x_i$ with $i = 1, 2, 3$ for the three directions.

What was said for temperature is equally true for any other quantity A:

$$\frac{dA}{dt} = \frac{\partial A}{\partial t} + \sum_i v_i \frac{\partial A}{\partial x_i}. \tag{1.60}$$

One speaks also of the Euler notation, working with $\partial/\partial t$, and of the Lagrange notation, working with the total derivative d/dt. Simple dots as symbols for derivatives with respect to time are dangerous in hydrodynamics.

If we now apply Newton's law of motion

$$\text{force} = \text{mass} \cdot \text{acceleration},$$

then the acceleration is the total time derivative of the velocity, since the particles of water are accelerating ("*substantial* derivative" dv/dt):

$$\text{force} = m\frac{dv}{dt} = m\left[\frac{\partial v}{\partial t} + (v\,\mathrm{grad})v\right].$$

Here $(v\,\mathrm{grad})A$ with a vector A means that $(v\,\mathrm{grad})$ is applied to each of the three components and that the result is a vector:

$$\left[(v\,\mathrm{grad})A\right]_k = \sum_i v_i \partial A_k/\partial x_i.$$

It is important to notice that the velocity v now occurs in Newton's law of motion not just linearly, but quadratically. Many problems in hydrodynamics accordingly are no longer soluble exactly for high velocities, but use up much calculation time on supercomputers. Clearly we measure dv/dt if we throw a scrap of paper into the stream and follow its acceleration; $\partial v/\partial t$ is being assessed if we hold a finger in the stream and feel the changing force on it. In both cases a bath-tub as a measuring environment is more practical than a bridge over the Mississippi.

Just as in the whole of continuum mechanics, we do not wish to consider the atoms individually, but to average them. We define therefore the density ρ as the ratio of mass to volume. More precisely ρ is the limiting value of the ratio of mass to volume, when the mass is determined in a notionally defined partial volume of the liquid, and this volume is very much greater than the volume of a single atom, but very much smaller than the total volume or the volume within which the density changes significantly. We take ρ simply to be the mass per cm^3, since the Mississippi is broader than a centimetre.

In a completely analogous manner we define the *force density* f as the force per cm^3 acting on a fluid (f = force/volume). Newton's law now has the form

$$f = \rho \left[\frac{\partial v}{\partial t} + (v \ \text{grad})v \right].$$ (1.61)

An example of the force density f is the force of gravity, $f = \rho g$. Later we shall also meet areal forces such as the pressure.

A "universally known" law is that of Gauss:

$$\oint j d^2 S = \int \text{div}(j) d^3 r$$ (1.62)

for a vector field $j = j(r)$. The left-hand side is a two-dimensional integral over the surface of the volume, over which the right-hand side is integrated three-dimensionally. The areal element $d^2 S$ is normal to this surface and points outwards. *Notation.* Two- or three-dimensional integrals, taken over a plane or a space, we denote by just an integral sign, and write the integration variable, for example, as $d^3 r$. An area integral, which extends, for example, over the closed surface of a three-dimensional volume, is denoted by an integral sign with a circle, as in (1.62); the area element is then a vector $d^2 S$, in contrast to $d^3 r$. In Stokes's law (1.81) will occur a closed one-dimensional line integral, which is also marked with a circle; these line integrals have a vector dl as integration variable pointing in the direction of the line. The notation dV for $d^3 r$ will be avoided here; in the section on heat the quantity V will be the magnitude of the volume in the mechanical work $-PdV$.

We now apply this calculation rule (1.62) to the current density $j = \rho v$ of the fluid stream; j thus represents how many grams of water flow per second through a cross-sectional area of one square centimetre, and points in the direction of the velocity v. Then the surface integral (1.62) is the difference between the outward and the inward flowing masses per second in the integration volume, and hence in the limit of a very small volume

$$-\partial(\text{mass})/\partial t = \text{div}(j) \cdot \text{volume}.$$

Accordingly after division by the volume we obtain the *equation of continuity*

$$\frac{\partial \rho}{\partial t} + \text{div}(j) = 0.$$ (1.63)

This fundamental relation between density variation and divergence of the relevant current density is valid similarly in many fields of physics, e.g., with electrical charge density and electrical current density. It is also familiar in connection with bank accounts: the divergence of outgoings and ingoings determines the growth of the reader's overdraft, and the growth in wealth of the textbook author.

A medium is called *incompressible* if its density ρ is constant:

$$\text{div}(\boldsymbol{j}) = 0, \quad \text{div}(\boldsymbol{v}) = 0. \tag{1.64}$$

Water is usually approximated as being *incompressible*, whereas air is rather compressible. Elastic solids also may be incompressible; then div $\boldsymbol{u} = 0$.

1.5.2 Stress, Strain and Hooke's Law

The force of gravity is, as mentioned, a volume force, which is measured by

$$\text{force density} = \text{force/volume}.$$

The pressure on the other hand has the dimension of force/area, and is therefore an areal force. In general we define an *areal force* as the limiting value of force/area for small area. Like force it is a vector, but the area itself can have various orientations. The areal force is therefore defined as a stress tensor σ:

σ_{ik} is the force (per unit area) in the i-direction

on an area at right-angles to the k-direction; $i, k = 1, 2, 3$. (1.65)

This tensor also is, like nearly all physical matrices, symmetric. Its diagonal elements σ_{ii} describe the pressure, which can indeed depend on the direction i in compressed solids; the non-diagonal elements such as σ_{12} describe the shear stresses. In liquids at rest the pressure P is equal in all directions, and there are no shear stresses: $\sigma_{ik} = -P\delta_{ik}$.

In the case when in a certain volume there is not only a volume force $\boldsymbol{f}$ but also an areal force σ acting on its surface, then the total force is

$$\boldsymbol{F} = \oint \sigma d^2 \boldsymbol{S} + \int \boldsymbol{f} d^3 r = \int (\text{div } \sigma + \boldsymbol{f}) d^3 r,$$

where we understand the divergence of a tensor to be the vector whose components are the divergences of the rows (or columns) of the tensor:

$$(\text{div } \sigma)_i = \sum_k \partial\sigma_{ik}/\partial x_k = \sum_k \partial\sigma_{ki}/\partial x_k.$$

In this sense we can apply Gauss's law (1.62) in the above formula. In the limiting case of small volume we therefore have

$$\frac{\text{areal force}}{\text{volume}} = \text{div } \sigma, \tag{1.66}$$

e.g., for the force which pressure differences exert on a cm^3.

Because of (1.66) the equation of motion now becomes

$$\rho\frac{d\boldsymbol{v}}{dt} = \text{div } \sigma + \boldsymbol{f} \tag{1.67}$$

with the total derivative according to (1.60), for solids as for liquids and gases. In a liquid at rest under the influence of gravity $\boldsymbol{f} = \rho\boldsymbol{g}$ we therefore have $\boldsymbol{f} = -\text{div } \sigma = \text{div}(P\delta_{ik}) = \text{grad } P$, and accordingly at height h: $P = \text{const.} - \rho gh$. For every ten metres of water depth the pressure increases by one "atmosphere" ≈ 1000 millibars $= 10^5$ pascals. Anybody who dives in the sea for treasure or coral must therefore surface very slowly, as the sudden lowering of pressure would allow dangerous bubbles to grow in the blood vessels. The relation div $\sigma = -\text{grad } P$ is also generally valid in "ideal" fluids without frictional effects (Euler 1755):

$$\rho\frac{d\boldsymbol{v}}{dt} = -\text{grad}(P) + \boldsymbol{f}. \tag{1.68}$$

Equation (1.68) gives three equations for four unknowns, $\boldsymbol{v}$ and ρ. If the flow is compressible we need also to know how the density depends on the pressure. As a rule we use a linear relation: $\rho(P) = \rho(P = 0)(1 + \kappa P)$, the compressibility κ being defined thereby.

In an elastic solid the *stress tensor* σ is no longer given by a unique pressure P, and instead of a unique compressibility we now need many elastic constants C. We again assume a linear relationship, only now between the stress tensor σ and the strain tensor ε,

$$\sigma = C\varepsilon, \tag{1.69}$$

analogous to Hooke's law: restoring force $= C\cdot$ displacement. Robert Hooke (1635–1703) would be somewhat surprised to be regarded as the father of (1.69), since σ_{ik} and ε_{mn} are indeed tensors (matrices). Consequently C is a tensor of the fourth order (the only one in this book), i.e. a quantity with four indices:

$$\sigma_{ik} = \sum_{mn} C_{mn}^{ik}\varepsilon_{mn} \quad (i, k, m, n = 1, 2, 3).$$

These 81 elements of the fourth order tensor C reduce to two Lamé constants μ and λ in isotropic solids:

$$\sigma = 2\mu\varepsilon + \lambda\mathcal{E}\mathrm{Tr}(\varepsilon) \tag{1.70}$$

with the unit matrix $\mathcal{E}$, and hence $\sigma_{ik} = 2\mu\varepsilon_{ik} + \lambda\delta_{ik}\Sigma_j\varepsilon_{jj}$. The compressibility is then (see Exercise) $\kappa = 3/(3\lambda + 2\mu)$, the ratio of pressure to relative change of length is the Young's modulus $E = \mu(2\mu + 3\lambda)/(\mu + \lambda)$. The ratio: relative change of length perpendicular to the direction of force divided by the relative change of length parallel to the direction of force is the Poisson's ratio $\lambda/(2\mu + 2\lambda)$. Accordingly, without proof, the elastic energy is given by $\Sigma_{ik}\mu(\varepsilon_{ik})^2 + (\lambda/2)(\mathrm{Tr}\varepsilon)^2$.

1.5.3 Waves in Isotropic Continua

Sound waves (long-wave acoustic phonons) propagate in air, water and solids with different velocities. How does it function? The mathematical treatment is the same in all cases, so long as frictional effects (acoustic damping) are ignored and we are dealing only with isotropic media, in which sound propagates with the same velocity in all directions. Then we have (1.70), but with $\mu = 0$, $\lambda = 1/\kappa$ for gases and liquids.

Acoustic vibrations have such small amplitudes (in contrast to shock waves) that they are treated in the harmonic approximation; quadratic terms such as $(v\,\mathrm{grad})v$ accordingly drop out: $dv/dt = \partial v/\partial t$. Therefore, taking account of (1.70), after some manipulation (1.67) takes the form

$$\rho\frac{\partial v}{\partial t} = \mathrm{div}\,\sigma + f = \mu\nabla^2 u + (\mu + \lambda)\mathrm{grad}\,\mathrm{div}\,u + f. \tag{1.71}$$

Here even for gases and liquids the displacement u makes sense, in that $v = \partial u/\partial t$, since all vibrations do indeed have a rest position. The Laplace operator ∇^2 is the scalar product of the nabla operator ∇ with itself:

$$\nabla^2 A = \nabla(\nabla A) = \mathrm{div}\,\mathrm{grad}\,A = \Sigma_i \partial^2 A/\partial x_i^2$$

for a scalar A. For a vector u, $\nabla^2 u$ is a vector with the three components $\nabla^2 u_1$, $\nabla^2 u_2$, $\nabla^2 u_3$. One should notice also the difference between div grad and grad div: operators are seldom commutative.

For the calculation of the sound velocity we neglect the gravity force f and assume sound propagation in the x-direction:

$$\rho\frac{\partial^2 u}{\partial t^2} = \mu\frac{\partial^2 u}{\partial x^2} + (\mu + \lambda)\mathrm{grad}\left(\frac{\partial u_x}{\partial x}\right) \tag{1.72}$$

or

$$\rho\partial^2 u_x/\partial t^2 = (2\mu + \lambda)\partial^2 u_x/\partial x^2, \quad \rho\partial^2 u_y/\partial t^2 = \mu\partial^2 u_y/\partial x^2,$$

in components, with the z-component analogous to the y-component. These equations have the form of the general *wave equation*

$$\frac{\partial^2 \Psi}{\partial t^2} = c^2 \frac{\partial^2 \Psi}{\partial x^2} \quad (\text{or} = c^2 \nabla^2 \Psi) \tag{1.73}$$

for the vibration Ψ, which for a plane wave has the solution

$$\Psi \sim e^{i(Qx - \omega t)} \quad \text{with} \quad \omega = cQ. \tag{1.74}$$

(For arbitrary direction of propagation Qx is to be replaced in the trial solution by $\boldsymbol{Q}\boldsymbol{r}$.) The sound velocity is given by $c = \omega/Q$, the velocity with which a definite phase is propagated, such as, for example, a zero of the real part $\cos(Qx - \omega t)$. (This phase velocity is to be distinguished from the group velocity $d\omega/dQ$, which may be smaller for high frequency phonons, but here coincides with ω/Q.) In three dimensions $\boldsymbol{Q}$ is the wave vector with magnitude $Q = 2\pi/(\text{wavelength})$; it is often denoted by $\boldsymbol{q}, \boldsymbol{k}$ or $\boldsymbol{K}$.

If we compare (1.73) with (1.72) in their three components we immediately see that

$$c^2 = (2\mu + \lambda)/\rho \tag{1.75}$$

for the case when the displacement u is parallel to the x-direction (longitudinal vibrations), and

$$c^2 = \mu/\rho \tag{1.76}$$

for transverse vibrations perpendicular to the x-direction. In general the sound is a superposition of longitudinal and transverse types of vibration. The longitudinal sound velocity is greater than the transverse velocity in solids, since in the longitudinal vibrations the density must also be compressed. In liquids and gases with $\mu = 0$ and $\lambda = 1/\kappa$ only longitudinal sound waves can exist (at low frequencies, as here assumed) with

$$c^2 = 1/(\kappa\rho). \tag{1.77}$$

Since the densities ρ can be different even in gases with the same compressibility κ, the sound velocity c always depends on the material. Usually one naturally thinks of sound in air under normal conditions.

1.5.4   Hydrodynamics

In this section we think less about solids but rather of isotropic liquids and gases. Nearly always we shall assume the flow to be incompressible, as suggested by water (hydro- comes from the Greek word for water).

(a) Bernoulli's Equation and Laplace's Equation

We call the flow static if $v = 0$, and steady if $\partial v/\partial t = 0$. (Is zero growth in the economy static or steady?) If the volume force f is conservative there is a potential ϕ with $f = -\text{grad}\,\phi$. Then in a steady incompressible flow with conservative volume force we have according to Euler's equation (1.68): $\rho(v\,\text{grad})v = -\text{grad}(\phi + P)$; here the pressure clearly becomes a sort of energy density (erg per cm^3).

Streamlines are the (averaged) velocity direction curves of the water molecules, and thus given mathematically by $dx/v_x = dy/v_y = dz/v_z$. If l is the length coordinate along a streamline, and $\partial/\partial l$ the derivative with respect to this coordinate in the direction of the streamline (hence in the direction of the velocity v), then we have $|(v\,\text{grad})v| = v\partial v/\partial l$, and hence for steady flows

$$-\partial(\phi + P)/\partial l = |-\text{grad}(\phi + P)| = \rho v\partial v/\partial l = \rho\partial(v^2/2)/\partial l$$

analogous to the derivative of the energy law in one dimension (see Sect. 1.1.3a). Along a steady streamline we therefore have

$$\phi + P + \rho v^2/2 = \text{const.} \tag{1.78}$$

(Bernoulli 1738). This is a conservation law for energy if one interprets the pressure, which derives from the forces between the molecules, as energy per cm^3; then ϕ is, for example, the gravitational energy and $\rho v^2/2$ the kinetic energy of a unit volume. This mechanical energy is therefore constant along a streamline, since friction is neglected. By measurement of the pressure difference one can then calculate the velocity.

A flow v is called a *potential flow* if there is a function Φ whose negative gradient is everywhere equal to the velocity v. Since quite generally the curl of a gradient is zero, for potential flows curl $v = 0$, i.e. the flow is "vortex free". If a potential flow is also incompressible, then we have $0 = \text{div}\,v = -\text{div grad}\Phi = -\nabla^2\Phi$ and

$$\nabla^2\Phi = 0 \quad \text{(Laplace Equation)}. \tag{1.79}$$

It can also be shown that (1.78) is then valid not only along a streamline, but also when comparing different streamlines:

$$\phi + P + \rho v^2/2 = \text{const.} \tag{1.80}$$

in the whole region of an incompressible steady flow without friction, with conservative forces.

(b) Vortex Flows

The well-known Stokes's law states that

$$\Gamma = \oint v dl = \iint \text{curl}(v)d^2S, \tag{1.81}$$

Fig. 1.17 Author's dream of
the Lorelei and her whirlpool
(schematic). Below and to
the right is seen a vortex:
velocity v as a function of
distance r from the vortex,
with core radius $a \to 0$

with the line integral dl along the rim of the area over which the integral d^2S is
integrated. In hydrodynamics Γ is called the circulation or *vortex strength*; it vanishes
in a potential flow. Since Thomson (1860) it is known that

$$\frac{d\Gamma}{dt} = 0, \tag{1.82}$$

for incompressible fluids without friction (even unsteady), i.e. the circulation moves
with the water particles.

With vortex lines, such as are realised approximately in a tornado, the streamlines
are circles about a vortex axis, similar to the force lines of the magnetic field round
a wire carrying a current. The velocity v of the flow is inversely proportional to the
distance from the vortex axis, as can be observed at the drain hole of a bath-tub. In
polar coordinates (r', ϕ) about the vortex axis an ideal vortex line therefore has the
velocity

$$v = e_\phi \Gamma / 2\pi r', \quad r' > a$$
$$v = e_\phi \omega r', \quad r' < a$$

with the core radius a and the angular velocity $\omega = \Gamma / 2\pi a^2$ within the core. Under
these conditions curl $v = 0$ outside the core and $= 2\omega$ in the core: the vortex strength
is concentrated almost like a point mass at the core, assumed small.

For a hurricane, the core is called the eye, and there it is relatively calm; the
film *Key Largo* is a good hurricane teacher. In the out-flowing bath-tub the core is
replaced by air. In modern physical research vortices are of interest no longer because
of the Lorelei, which enticed the Rhine boatmen into the whirlpool of old (Fig. 1.17),
but because of the infinitely long lifetime of vortices in superfluid helium[3] at low
temperatures because of quantum effects (Onsager, Feynman, around 1950). Also
the lift of an aeroplane wing arises from the circulation about the wing; the wing is
therefore the core of a sort of vortex line.

If two or more vortex lines are parallel side by side in the fluid, the core of each
vortex line must move in the velocity field arising from the other vortex lines. For

---

[3] E.L. Andronikashvili and Yu.G. Mamaladze, p. 79 in: Progress in Low Temperature Physics, vol. V,
edited by C.J. Gorter (North Holland, Amsterdam 1967).

Fig. 1.18 Motion of a vortex pair with equal (*left*) and opposite (*right*) circulations

Fig. 1.19 Flow between two plates (side view) for determination of the viscosity η. The *upper plate* moves with velocity v_0 to the *right*, in contrast to the *lower plate*, which is fixed

the circulation is concentrated on a thin core and must move with the fluid, as stated above. So two parallel vortex filaments with $\Gamma_1 = -\Gamma_2$ follow a straight line course side by side, whereas if $\Gamma_1 = +\Gamma_2$ they dance round each other (Fig. 1.18). If one bends a vortex line into a closed ring, then for similar reasons this vortex ring moves with unchanging shape in a straight line: each part of the ring must move in the velocity field of all the other pans. This vorticity is also the reason why one can blow out candles, but not suck them out (danger of burning in proving this experimentally!). Also experienced smokers can blow smoke-rings (if the non-smokers let them).

(c) Fluids with Friction

In the "ideal" fluids investigated up to this point there is no friction, and so the stress tensor σ consists only of the pressure P: $\sigma_{ik} = -P\delta_{ik}$. If, however, we stir honey with a spoon we create shear stresses such as σ_{12}, which are proportional to the velocity differences.

Just as two elastic constants μ and λ sufficed in the elasticity theory for isotropic solids in (1.70), we need only two *viscosities*, η and ζ (with $\mathcal{E}$ = unit tensor) for the stresses caused by friction:

$$\sigma' = 2\eta\varepsilon' + (\zeta - 2\eta/3)\mathcal{E}\,\mathrm{Tr}(\varepsilon'). \tag{1.83}$$

Here σ' is the stress tensor without the pressure term, and ε' has the matrix elements $(\partial v_i/\partial x_k + \partial v_k/dx_i)/2$, since the corresponding expression with u in (1.59) makes little sense for fluids. The trace of the tensor ε' is then simply div v, so that in incompressible flows the complicated second term in (1.83) drops out. Thus hydrodynamics usually requires only one friction coefficient whereas elasticity theory needs two.

Let us consider as an example the flow between two parallel plates perpendicular to the z-axis (Fig. 1.19). The upper plate at $z = d$ moves with velocity v_0 to the right, the lower plate at $z = 0$ is fixed. After some time a steady fluid flow is established between the plates: v points only in the x-direction to the right, with $v_x(z) = v_0 z/d$, independently of x and y. Accordingly div $v = 0$: the flow is incompressible then, even if the fluid itself is compressible. The tensor $\sigma' = 2\eta\varepsilon'$ according to (1.83) contains many zeros, since only $\varepsilon'_{13} = \varepsilon'_{31} = (\partial v_x/\partial z + 0)/2 = v_0/2d$ is different from zero:

$$\sigma'_{13} = \eta v_0/d.$$

This is therefore the force in the x-direction, which is exerted on each square centimetre of the plates perpendicular to the z-direction, in order to overcome the frictional resistance of the fluid. In principle the viscosity η can be measured in this way, although falling spheres (see below) are a more practical method for determining the viscosity. The other viscosity ζ only comes into it if the density changes, as for example in the damping of shock waves.

With this stress tensor σ' and the pressure P, (1.67) has the form

$$\rho d v/d t = \text{div } \sigma' - \text{grad } P + f,$$

which can be rewritten (see (1.83)) to be analogous to (1.71)

$$\rho\frac{dv}{dt} = \eta\nabla^2 v + (\zeta + \eta/3)\text{grad div}v - \text{grad } P + f. \tag{1.84}$$

In the special case of incompressible flow div $v = 0$ and $f = 0$ this yields the celebrated *Navier-Stokes* equation (1822):

$$\rho\frac{dv}{dt} = \eta\nabla^2 v - \text{grad } P, \tag{1.85}$$

which has already used up much work and storage in many computers. Since ρ is now constant we can, if the pressure is also constant, define the kinematic viscosity $\nu = \eta/\rho$, and write

$$\frac{dv}{dt} = \nu\nabla^2 v. \tag{1.86}$$

This equation has the form of a *diffusion* or heat conduction equation, ignoring the difference (negligible for small velocities) between dv/dt and $\partial v/\partial t$. A high velocity concentrated in one place is therefore propagated outwards by friction just like the temperature of a solid briefly heated in one place, until eventually the whole fluid has the same velocity. The solution is $\exp(-t/\tau)\sin(Qr)$ with $1/\tau = \nu Q^2$, if $\sin(Qr)$ is the starting condition, whether it is the propagation of small velocities in a viscous fluid, heat in a solid, or molecules in a porous material. In air, water and glycerine ν is of order 10^{-1}, 10^{-2} and $10\,\text{cm}^2/\text{s}$, respectively.

Fig. 1.20 Poiseuille flow through a long tube, with parabolic velocity profile $v_x(r), 0 < r < R$

(d) Poiseuille Law (1839)

A somewhat more complicated flow than that described above between moving and fixed plates is that through a long tube (Fig. 1.20). In the middle the water flows fastest, at the walls it "sticks". For the steady solution we require the Navier-Stokes equation: $0 = -\text{grad } P + \eta \nabla^2 v$, or since all the flow is only towards the right in the x-direction: $\partial P / \partial x = \eta \nabla^2 v_x$. P is independent of y and z, whereas v_x is a function of the distance r from the centre of the tube; $v_x(r = R) = 0$ at the wall of the tube with radius R.

For a quantity A independent of angle we have in general

$$\nabla^2 A = d^2 A/dr^2 + \frac{d-1}{r} dA/dr$$

in d dimensions. Here $d = 2$ (polar coordinates for the cross-section of the tube); moreover $P' = \partial P / \partial x = -\Delta P / L$ in a tube with length L and pressure difference ΔP. Accordingly, we have to solve

$$P' = \eta \left(d^2 v_x/dr^2 + \frac{1}{r} dv_x/dr \right) = \frac{\eta}{r} d(r dv_x/dr)/dr,$$

(these transformations of ∇^2 are also useful elsewhere). We find that

$$r dv_x/dr = P'r^2/2\eta + \text{const.},$$
$$v_x = P'r^2/4\eta + \text{const } \ln(r) + \text{const.}'.$$

Since the velocity at $r = 0$ must be finite, the const is zero, and since at $r = R$ the velocity must be zero, $\text{const.}' = -P'R^2/4\eta$, so that

$$v_x = \frac{\Delta P}{4L\eta}(R^2 - r^2), \tag{1.87}$$

and the velocity profile is a parabola. The total flow through the tube (grams per second) is

$$J = \rho \iint v_x(r) dz dy = (\rho \Delta P \pi/8L\eta) R^4$$

so that

$$J \sim R^4. \tag{1.88}$$

The flow of water through a tube is therefore not proportional to the cross-section, but to the square of the cross-section, since the maximal velocity in the centre of the tube, (1.87), is itself proportional to the cross-section. This law also can be applied to the measurement of the viscosity. It no longer holds when the steady flow becomes unstable at high velocities because of turbulence.

Modern research in hydrodynamics has to do with, for example, the flow of oil and water through porous media. When an oil well "runs dry" there is still a great deal of oil in the porous sand. When one tries to squeeze it out by pumping water into the sand, complex instabilities arise, with beautiful, but unhelpful, fractal structures (see Chap. 5). Hydrodynamics is no dead formalism!

Fractal[4] is the name given to objects with masses proportional to (radius)^D and a fractal dimension D differing from the space dimension d; other fractals are snowflakes, the path of a diffusing particle, polymer chains in solutions, geographical features, and also the "clusters" which the magnetism program of Sect. 2.2.2 produces on the computer near the Curie point. Since about 1980, fractals (see Chap. 5) have been a rapidly expanding research field in physics.

(e) Similarity Laws

Quite generally, one should always try first to solve complicated differential equations in dimensionless form. Thus, if one divides all velocities by a velocity typical of the flow v_0, all lengths by a typical length l, etc., setting $r/l = r'$, $v/v_0 = v'$, $t/(l/v_0) = t'$, $P/(\rho v_0^2) = P'$ then (1.85) takes the dimensionless form

$$\mathrm{d}\boldsymbol{v}'/\mathrm{d}t' = \nabla'^2 \boldsymbol{v}'/\mathrm{Re} - \mathrm{grad}' \, P',$$

where Re is the so-called *Reynolds number*, defined as

$$\mathrm{Re} = v_0 l \rho / \eta = v_0 l / \nu \tag{1.89}$$

We can study the Navier-Stokes equation without knowing v_0 and l; one only needs to know the value of Re. If one has found a solution (exact, on the computer, or by experiment) of the Navier-Stokes equation for a certain geometry, the flow for a similar geometry (uniform magnification or shrinking factor) is similar, if only the Reynolds number is the same. A tanker captain can therefore get initial experience in the control of a ship in a small tank, if the flow conditions in the tank reproduce the full-scale flow with the same Reynolds number (if we neglect gravity).

It turns out, for example, that the steady solutions obtained so far are stable only up to Reynolds numbers of about 10^3. Above that value turbulence sets in, with the spontaneous formation of vortices. This also is a current field of research.

---

[4]See, e.g., B. Mandelbrot: The Fractal Geometry of Nature (Freeman, New York, San Francisco 1982); also Physica D 38 (1989).

If, for example, one heats a flow between two plates from below, "Rayleigh-Benard" instabilities occur with large temperature differences Δ, and these are also observed in the atmosphere (spacewise periodic clouds). With particularly large Δ the heat flow increases with an experimentally determined $\Delta^{1.28}$ (Libchaber and co-workers 1988) in contrast to normal heat conduction; theoretically an exponent 9/7 is predicted.

If a sphere of radius R sinks under its own weight through a viscous fluid with velocity v_0, then the ratio: force/$(\rho v_0^2 R^2)$ is dimensionless and therefore according to the Navier-Stokes law is a function only of the Reynolds number $\text{Re} = v_0 R / \nu$. For small Re this frictional force F is proportional as usual to the velocity: $F = \text{const.}(\rho v_0^2 R^2)/\text{Re} = \text{const.} v_0 R \eta$. Exact calculation gives const. $= 6\pi$ and hence the Stokes law

$$F = 6\pi \eta v_0 R. \tag{1.90}$$

Our dimensional analysis has thus spared us much calculation, but of course does not provide the numerical factor 6π. The Stokes law provides a convenient method for measuring η.

Another dimensionless ratio is the Knudsen number $\text{Kn} = \lambda/l$, where λ is the mean free path length of gas molecules. Our hydrodynamics is valid only for small Knudsen numbers. Other examples are the Peclet number, the Nusselt number and the Rayleigh number.

In conclusion it should be noticed that the forces acting on solids, liquids and gases, such as we have been treating here, are quite generally linked by linear combinations of the tensors ε and σ, their traces and their time derivatives. Our results up to now are therefore special cases: our simple equation $\rho(P) = \rho(P = 0)(1 + \kappa P)$ uses only $\text{Tr}(\sigma)$ and $\text{Tr}(\varepsilon)$: the much more complicated equation (1.69) links σ and ε and (1.83) also does this (only ε is then defined by the time derivative of the position).

Questions

Section 1.1

1. State Kepler's third law.
2. When do force-free bodies move in a straight line?
3. What force does a stone exert on a string when it is whirled round at constant speed?
4. With what speed must we throw a stone upwards, in order that it should escape the earth's gravity field? (Energy conservation: potential energy is $-GMm/r$, where r is the distance from the centre of the earth.)
5. Estimate the numerical value of the mean density of the earth ρ, from G, g and the earth's radius R.

Section 1.2

6. What is the "reduced mass" in the two-body problem?
7. State d'Alembert's Principle on constraining forces.
8. State the principle of virtual displacement with constraining forces.

Section 1.3

9. Why does Hamilton's principle apply only with fixed endpoints?
10. What are the variables of the Lagrange function L, and those of the Hamilton function H?
11. What are optic and acoustic phonons?

Section 1.4

12. What are the relationships between torque M, angular momentum L, inertia tensor Θ and angular velocity ω? Is ω a vector?
13. What are the "principal axes" of an inertia tensor, and what are the (principal) moments of inertia?
14. What is the nutation frequency of a cube rotating about an axis of symmetry?
15. Do the Euler equations determine the amplitude of the nutation of a symmetrical gyroscope?
16. Why does the axis of the gyroscope move perpendicularly to the applied force?
17. What is "Larmor precession" and what is it used for?

Section 1.5

18. What is the difference between $\partial/\partial t$ and d/dt in continuum physics?
19. What is an equation of continuity?
20. What are the relationships between pressure, stress tensor and strain tensor?
21. What is the difference between: hurricane, typhoon and tornado?
22. What is the meaning of: incompressible, vortex-free, ideal, steady, static?
23. For what values of the "Knudsen number" λ/R is Stokes's formula for the motion in air of spheres (radius R) valid?
24. With what diffusion constant D does a cluster of spheres disperse in a viscous fluid, when according to Einstein diffusivity/mobility $= k_B T$?

Problems

Section 1.1

1. Is a uniform motion in a straight line transformed into uniform motion in a straight line by a Galileo transformation?
2. Describe in one or two pages the Coriolis force, e.g., when shooting polar bears at the north pole.
3. A point mass moves on a circular orbit round an isotropic force centre with potential $\sim r^{-x}$. For what values of x is this orbit stable, i.e., at a minimum of the effective potential energy?
4. With what velocity does a point mass fall from the height h to earth, first if $h \ll$ earth radius, then generally?

Section 1.3

5. Using the principle of virtual displacements, calculate the pressure on the piston, if the force F acts on the wheel.

6. Lagrange equation of the first kind in cylindrical coordinates: a point mass moves in a gravity field on a rotationally symmetric tube $\rho = W(z)$, with $\rho^2 = x^2 + y^2$, where the height h and the angular velocity ω are constant ("centrifuge"). What shape $W(z)$ must the tube have, if ω is to be independent of z? *Hint*: resolve the acceleration into components e_r, e_ϕ and e_z in cylindrical coordinates.

7. Study the Lagrange equation for a thin hoop rolling down a hillside.
   Hint: $T = T_{\text{trans}} + T_{\text{rot}}$; all point masses are at the same distance from the centre.

8. Prove in the general case that $\{q_\mu, p_\nu\} = \delta_{\nu}$, $\{p_\mu, p_\nu\} = \{q_\mu, q_\nu\} = 0$, $\{F_1 F_2, G\} = F_1\{F_2, G\} + F_2\{F_1, G\}$ and hence study $d\boldsymbol{p}/dt = \{\boldsymbol{p}, H\}$ for the three-dimensional oscillator, $U = Kr^2/2$.

9. Using the harmonic approximation calculate the vibration frequencies of a particle in a two-dimensional potential $U(x, y)$.

Section 1.4

10. Discuss in the harmonic approximation the stability of free rotation of a rigid body about its principal axes, with $\Theta_1 > \Theta_2 > \Theta_3$.

11. Which properties of matrices does one need in theoretical mechanics? Is $(\Theta_1, \Theta_2, \Theta_3)$ a vector? Is $\Theta_1 + \Theta_2 + \Theta_3$ a "scalar", i.e., invariant under rotation of the coordinate axes?

12. Calculate the inertia tensor of a cylinder with mass M, radius R and height H in a convenient system of reference.
    Hint: $\int_0^1 (1 - x^2)dx = \pi/4$ and $\int_0^1 (1 - x^2)^{3/2}dx = 3\pi/16$.

Section 1.5

13. What is the form of the strain tensor if a specimen expands in the x-direction by one part in a thousand, shrinks in the y-direction by one part in a thousand, and stays unchanged in the z-direction? What is the volume change? What is changed if we also have $\varepsilon_{13} = 10^{-3}$? What follows generally from $\varepsilon_{ki} = \varepsilon_{ik}$?

14. An iron wire is extended by a tension ΔP (= force/area). Prove that $\Delta P = (\Delta l/l)E$ for the change in length, with $E = (2\mu + 3\lambda)\mu/(\mu + \lambda)$, and $\Delta V/V = \Delta P\kappa/3$ for the volume change, with $\kappa = 3/(2\mu + 3\lambda)$.

15. Use the Gaussian theorem $\nabla^2 1/r = -4\pi\delta(r)$ to show that a vortex line is (almost) a potential flow. At what speed do two vortices with the same circulation move around each other?

Chapter 2
Electricity and Magnetism

Abstract From Coulomb's law force = charge$_1 \times$ charge$_2$/distance^2 to Einstein's relativistic electrodynamics of moving bodies, via Maxwell equations with and without matter, we need to get to this chapter's end to understand why the above equality is not merely a proportionality, in our CGS units: Magnetic fields are the relativistic corrections to the electric fields.

An electric charge at rest gives rise to an electric field, a moving charge causes additionally a magnetic field, and an oscillating charge causes electromagnetic waves. How can one describe this theoretically and achieve a better understanding through the theory of relativity? First of all we shall treat an individual point charge in a vacuum, then the behaviour of matter, and finally Einstein's Theory of Relativity. We are working here with a system of units in which the electric field E and the magnetic field B have the same units, since relativistically they are only the various components of an antisymmetric 4×4 field tensor.

2.1 Vacuum Electrodynamics

2.1.1 Steady Electric and Magnetic Fields

Experience shows that there are other forces than gravitation. Here we consider the electromagnetic force $F = q(E + (v/c) \times B)$ on an electric charge moving with velocity v; c is the velocity of light. At first we shall ignore the magnetic field B.

(a) Coulomb's Law.
Between two stationary electric charges q_1 and q_2 at a distance r apart there is an isotropic central force F, falling off as $1/r^2$:

$$F = \text{const.} \, q_1 q_2 e_r / r^2. \qquad (2.1)$$

If the unit of measurement for the charge is already fixed, the proportionality factor has to be determined experimentally, e.g., by $1/\text{const.} = 4\pi\varepsilon_0$. Theoretical physicists make life easy by setting: const. $= 1$ in the cgs system. The unit of charge esu

© Springer-Verlag GmbH Germany 2017
D. Stauffer et al., *From Newton to Mandelbrot*, Graduate Texts in Physics,
DOI 10.1007/978-3-662-53685-8_2

(electrostatic unit) therefore repels another unit charge at a distance of 1 cm with a force of 1 dyne $= 1$ cm g s^{-2}. The SI unit "1 coulomb" corresponds to 3×10^9 esu; an electron has a charge of -1.6×10^{-19} coulombs or -4.8×10^{-10} esu. If one coulomb per second flows along a wire, then that is a current of 1 ampere; if it falls through a potential drop of one volt, then the power is 1 watt = 1 volt $\times$ ampere and the work done in one second is 1 watt $\times$ second = 1 joule. For technical applications, therefore, SI units are more practical than our cgs units.

The difference between the SI system, with ampere as a current unit besides meter, kilogram and seconds, and the cgs system without a separate electric unit and only centimeter, gram and second, still arises big emotions. Colleagues at Cologne University cited a remark by Val L. Fitch, 1980 Nobel prize in physics: "any system that gives E and F different units, when they are related through a relativistic transformation, is on the far side of sanity" (1999). Unfortunately, this wisdom gets clear only at the end of this chapter in Sect. 2.3.2 on relativistic electrodynamics; until then regard our choice of cgs just as laziness to avoid writing ε_0 and μ_0 in numerous formulas. On the other hand, if you dislike this cgs system where a capacitance is measured in centimeters, read the old French translation of this book.

For an electron and a proton their Coulomb attraction is 10^{39} times as strong as their attraction by gravitation. The fact that we nevertheless detect the force of gravitation depends on the fact that there are positive and negative charges, but only positive masses. The electrostatic forces which hold the atoms together therefore cancel out, when seen externally, leaving only the usual gravitational effect, causing the earth to orbit round the sun.

The field strength E is defined by the force on a positive unit charge: $F = qE$. Equation (2.1) then becomes

$$E = (q/r^2)e_r \tag{2.2}$$

for the field about a point charge q. It is usually more practical, as in mechanics, to work with a potential energy, since Coulomb forces are conservative. In this sense the *potential ϕ* is defined as the potential energy of a unit charge, hence $E = -\text{grad}\,\phi$. The *potential difference U* (SI unit: volt) is the difference in potential between two points. So the Coulomb potential for a charge q is

$$\phi = q/r \tag{2.3}$$

The fields E arising from different point charges are linearly superposed:

$$E = \sum_i q_i e_i/r_i^2 = \int \left[\rho(r)e_r/r^2 \right] d^3r$$

for the field at the origin of coordinates, with charge density ρ (esu per cm^3), or

$$\phi = \int \frac{\rho(r)}{r} d^3r.$$

To obtain the field or the potential at an arbitrary position r one only has to replace r by the distance:

$$\phi(r) = \int \frac{\rho(r')}{|r - r'|} d^3 r'. \tag{2.4}$$

A *delta function* $\delta(x)$ (delta "distribution" in the formal jargon) is a very high and narrow peak, such as, for example, the density of a point mass, and is the generalisation of the Kronecker delta δ_{ij} to real numbers: $\delta(x) = 0$ except when $x = 0$; the integral over the delta function is unity, and for any function $f(x)$ we therefore have

$$\int_{-\infty}^{\infty} f(x)\delta(x)dx = f(0). \tag{2.5}$$

The analog holds in three dimensions, with $\delta(r) = \delta(x)\delta(y)\delta(z)$.

Moreover, $\nabla^2 r^{-1} = \text{div grad}(1/r) = -4\pi\delta(r)$, as may be proved using Gauss's law:

$$\int \text{div grad}(1/r)d^3 r = \oint \nabla r^{-1} d^2 S = 4\pi r^2/(-r^2) = -4\pi.$$

For the electrostatic potential and field we therefore have

$$\text{div } E = -\text{div grad } \varphi = \int -\nabla_r^2 \frac{\rho(r')}{|r - r'|} d^3 r'$$

$$= 4\pi \int \rho(r')\delta(r - r')d^3 r' = 4\pi\rho(r).$$

In general the curl of a gradient is zero: since $E = -\text{grad}\phi$ we accordingly have:

$$\text{div } E = 4\pi\rho, \quad \text{curl } E = 0. \tag{2.6}$$

Using the laws of Gauss and Stokes one can also write this in integral form:

$$\oint E d^2 S = 4\pi Q, \quad \text{and} \quad \oint E dl = 0 \text{ with } Q = \int \rho(r)d^3 r.$$

This primitive form of Maxwell's equations is valid only for charges at rest: "electrostatic". Figure 2.1 shows the definition of this surface integral and this line integral.

If the charge density is zero, then div $E = 0$, so since div grad $= \nabla^2$ we obtain the Laplace equation $\nabla^2\phi = 0$ for the potential. The potential at the boundary can be laid down by external voltages. The partial differential equation $\nabla^2\phi = 0$ can often be solved exactly in two dimensions; in other cases one does it using the computer (Program LAPLACE). To this end we divide up the space under investigation into individual cubes i, j, k, which are each described by one point $\phi i, j, k$ (i, j, k are integers). Replacing the differential quotients in ∇^2 by divided differences, we reduce $\nabla^2\phi = 0$ in two dimensions to the formula

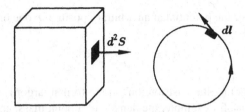

Fig. 2.1 The surface integral of E (*left*) gives the charge enclosed in the volume. The line integral of E (*right*) is zero. The surface element d^2S and the line element dl are vectors pointing outwards and in the direction of the integration, respectively

$$\phi(i, k) = [\phi(i + 1, k) + \phi(i - 1, k) + \phi(i, k + 1) + \phi(i, k - 1)]/4.$$

One solves this iteratively, i.e. starting from a trial solution ϕ one calculates for each point the right-hand side and then replaces the old ϕ at this point (i, k) by this result. This substitution is repeated until ϕ is scarcely changing any more. The program prints out ϕ along a straight line in the middle of a square, if ϕ is fixed on the perimeter of a square as 0 or 1; the trial initial solution is $\frac{1}{2}$. Loops of the type "for $i = 1$ to 20" execute all the following instructions as far as "next i"; they can also be nested within each other.

PROGRAM LAPLACE

```
10 dim ph(20,20)
20 for i=1 to 20
30 for k=1 to 20
40 ph(i,k)=0.5
50 next k
60 ph(i,1)=0
70 ph(i,20)=0
80 next i
90 for k=1 to 20
100 ph(1,k)=0
110 ph(20,k)=0
120 next k
130 print 1,ph(1,10)
140 for i=2 to 19
150 for k=2 to 19
160 ph(i,k)=0.25*(ph(i-1,k)+ph(i+1,k)+ph(i,k-1)+ph(i,k+1))
170 next k
180 next i
190 for i=1 to 20
200 print i,ph(i,10)
210 next i
220 print
230 goto 140
240 end
```

(b) Magnetic Fields

Moving charges are subject also to additional "*Lorentz*" forces perpendicular to the velocity v and proportional to the electrical charge q. The proportionality factor is called the *magnetic field* B'; here we measure v in units of the velocity of light c:

$$F = (q/c)v \times B. \tag{2.7}$$

The sudden occurrence here of an absolute velocity v (relative to the reader's desk?) already indicates the need for a relativistic interpretation (Sect. 2.3). The continuity equation (1.63) is also valid here for electrical charge density ρ and current density j:

$$\frac{\partial \rho}{\partial t} + \text{div } j = 0. \tag{2.8}$$

Individual magnetic charges ("monopoles") have not so far been discovered; if one breaks a bar magnet in the middle, one does not obtain separately a north pole and a south pole, but two "*dipoles*", each with a north pole and a south pole. The magnetic charge density is therefore always zero; analogously to div $E = 4\pi \rho$ we have div $B = 0$. B is caused not by charges, but by currents:

$$\text{curl } B = 4\pi/c j. \tag{2.9}$$

For example, at a distance r from a wire carrying a current the magnetic field is $B = 2J/cr$, where J is the area integral over the current density j, and hence the total current in the wire. For proof one only needs to carry out this area integral over (2.9) and apply Stokes's law to the left-hand side:

$$\iint \text{curl } B d^2 S = \oint B dl = 2\pi r B.$$

(Instead of the factor $4\pi/c$ in (2.9) one needs other proportionality factors if one is using other systems of units.)

In general a vector field $F(r)$ can be constructed from its divergence and its curl:

$$F(r) = \int \frac{(\text{div } F)R + (\text{curl } F) \times R}{4\pi R^3} d^3 r' + F_{\text{hom}}(r) \tag{2.10}$$

with $R = r - r'$. Here F_{hom} is a solution of div $F = $ curl $F = 0$, compatible with the boundary conditions, hence, for example, $F_{\text{hom}} = $ const. This is in effect the three-dimensional generalisation of

$$F(x) = \int (dF/dx) dx + \text{const.}$$

in one dimension. For the magnetic field from an arbitrary finite current distribution $j(r)$ this yields the Biot-Savart law

$$\boldsymbol{B}(\boldsymbol{r})c = \int \left[\boldsymbol{j}(\boldsymbol{r}') \times (\boldsymbol{r} - \boldsymbol{r}') \right] / |\boldsymbol{r} - \boldsymbol{r}'|^3 \mathrm{d}^3 r',$$

if $\boldsymbol{B}$ is to vanish at infinity. In spite of their complexity, however, these equations are still not complete and do not explain, for example, why a transformer works with alternating current and not with direct current. The results obtained up to this point are in fact valid only for the steady case, where no currents or fields change with time.

2.1.2 Maxwell's Equations and Vector Potential

(a) Maxwell's Equations
It can be shown experimentally that a magnetic field $\boldsymbol{B}$ which changes with time induces potential differences U in a fixed wire loop according to the law of induction:

$$cU = \int\!\!\int (\partial B / \partial t) \mathrm{d}^2 S.$$

Since, on the other hand,

$$U = -\oint \boldsymbol{E}(\boldsymbol{l})\mathrm{d}\boldsymbol{l} = -\int\!\!\int \operatorname{curl} \boldsymbol{E} \mathrm{d}^2 S,$$

we deduce that $c \operatorname{curl} \boldsymbol{E} = -\partial \boldsymbol{B} / \partial t$. The previously obtained result $c \operatorname{curl} \boldsymbol{B} = 4\pi \boldsymbol{j}$ also must be generalised, in order that the continuity equation (1.64) should still hold for ρ varying with time: to $\boldsymbol{j}$ must be added Maxwell's displacement current $(\partial \boldsymbol{E} / \partial t)/4\pi$. Now we have the four equations of Maxwell complete:

$$\operatorname{div} \boldsymbol{E} = 4\pi\rho, \ \ c \operatorname{curl} \boldsymbol{E} + \frac{\partial \boldsymbol{B}}{\partial t} = 0, \ \ \operatorname{div} \boldsymbol{B} = 0, \ \ c \operatorname{curl} \boldsymbol{B} - \frac{\partial \boldsymbol{E}}{\partial t} = 4\pi \boldsymbol{j}. \quad (2.11)$$

together with the general Lorentz force $\boldsymbol{F}/q = \boldsymbol{E} + \boldsymbol{v} \times \boldsymbol{B}/c$. Experience shows that these equations of Maxwell are correct and are known in students' examinations. The theory of relativity does not require any more changes and only makes them clearer. We have here the model of a successful coherent theory for a large number of apparently separate phenomena, analogous to Newton's law of motion in mechanics or Schrodinger's equation in quantum mechanics. The modern universal formulae, describing the forces between the elementary particles, are unfortunately more complicated.

(b) Vector Potential

We have defined $E = -\text{grad}\,\phi$ in electrostatics. Now we wish to generalise this, and also obtain something similar for B. Since curl B is non-zero even in the steady case, B, unlike E, can hardly be the gradient of a potential.

Instead, we define for the magnetic phenomena a "*vector potential*" A, such that

$$\text{curl}\,A = B.$$

Since curl grad $f(r) = 0$, $A + \text{grad}\,f(r)$, with an "arbitrary" function f, is a vector potential leading to the same B as does A. One of Maxwell's equations now becomes

$$0 = c\,\text{curl}\,E + \frac{\partial B}{\partial t} = \text{curl}\left(cE + \frac{\partial A}{\partial t}\right);$$

and since this curl is zero, this suggests replacing the earlier definition $E = -\text{grad}\,\phi$ by

$$E + (1/c)\frac{\partial A}{\partial t} = -\text{grad}\,\phi$$

giving:

$$B = \text{curl}\,A \quad \text{and} \quad E + \dot{A}/c = -\text{grad}\,\phi. \tag{2.12}$$

As integration constants we set $\phi = A = 0$ at infinity, together with either the "Coulomb gauge" div $A = 0$, or rather the "Lorentz gauge":

$$c\,\text{div}\,A + \frac{\partial \phi}{\partial t} = 0. \tag{2.13}$$

The above Biot-Savart law for steady currents $j(r)$ now becomes simpler:

$$A = \int \left[j(r')/c|r - r'|\right]\,\mathrm{d}^3 r'.$$

In general one can deduce from the Maxwell equations and the Lorentz gauge (with curl curl $A = \text{grad}\,\text{div}A - \nabla^2 A$)

$$\Box A + (4\pi/c)j = 0 \quad \text{and} \quad \Box\phi + 4\pi\rho = 0; \quad \text{with} \quad \Box = \nabla^2 - c^{-2}\partial^2/\partial t^2 \tag{2.14}$$

as the wave operator (d'Alembert operator, "quabla"). Wave equations $\Box f = 0$ are solved by propagating waves: $f(x, t) = F(x \pm ct)$ in one dimension, with an arbitrary shape F for the wave profile. Playing with a rope fastened at one end might have provided the reader's first experiments with (2.14).

More serious is the fact that the fields E and B now have a life of their own: even without charges ρ and currents j electromagnetic waves are possible in consequence of $\Box A = \Box\phi = 0$. Theoretical physicists take pride in the fact that these waves were predicted *before* Heinrich Hertz demonstrated them in the laboratory more than a hundred years ago. These waves have been "seen" naturally since time immemorial, as light is just a superposition of such electromagnetic waves.

2.1.3   Energy Density of the Field

If in hydrodynamics a mass were to fall "from the sky", then the equation of continuity would have the form: $\partial\rho/dt + \operatorname{div} \boldsymbol{j} = $ influx from outside. If now in electrodynamics we derive an equation of the form $\partial u/\partial t + \operatorname{div} \boldsymbol{S} = $ electric power density, then we shall recognise that u is the energy density and $\boldsymbol{S}$ is the energy flow density. We obtain such a form when we multiply two of Maxwell's equations scalarly by $\boldsymbol{E}$ and $\boldsymbol{B}$:

$$\boldsymbol{B} \operatorname{curl} \boldsymbol{E} + (\partial B^2/\partial t)/2c = 0,$$
$$\boldsymbol{E} \operatorname{curl} \boldsymbol{B} - (\partial E^2/\partial t)/2c = (4\pi/c)\boldsymbol{j}\boldsymbol{E}.$$

The difference between the two equations gives

$$2c(\boldsymbol{B} \operatorname{curl} \boldsymbol{E} - \boldsymbol{E} \operatorname{curl} \boldsymbol{B}) + \partial(E^2 + B^2)/\partial t = -8\pi \boldsymbol{j}\boldsymbol{E}.$$

On the right-hand side $-\boldsymbol{j}\boldsymbol{E} = -\rho\boldsymbol{v}\boldsymbol{E}$ is the power density (force times velocity per unit volume), and hence the measure of electric energy per unit time transformed by resistance ("friction") into heat, and so the field energy u lost. On the left-hand side: $\boldsymbol{B} \operatorname{curl} \boldsymbol{E} - \boldsymbol{E} \operatorname{curl} \boldsymbol{B} = \operatorname{div}(\boldsymbol{E} \times \boldsymbol{B})$. We have therefore derived the desired form for energy density u and energy flow density $\boldsymbol{S}$:

$$u = (E^2 + B^2)/8\pi; \quad \boldsymbol{S} = (c/4\pi)\boldsymbol{E} \times \boldsymbol{B} \quad \text{(Poynting vector)}. \tag{2.15}$$

Just as the Hamilton function $p^2/2m + Kx^2/2$ of the harmonic oscillator leads to an energy exchange between kinetic and potential energies, the energy density $u \sim E^2 + B^2$ gives the possibility of electromagnetic waves, where electrical and magnetic energies are exchanged Moreover, $\boldsymbol{S}/c^2$ is the momentum density, which remains fixed in the wave field.

2.1.4   Electromagnetic Waves

Even without a vector potential the Maxwell equations lead to the wave equation *in vacuo* ($\rho = \boldsymbol{j} = 0$):

$$(1/c^2)(\partial^2\boldsymbol{E}/\partial t^2) = (1/c)\partial(\operatorname{curl} \boldsymbol{B})/\partial t = (1/c)\operatorname{curl}(\partial\boldsymbol{B}/\partial t)$$
$$= -\operatorname{curl} \operatorname{curl} \boldsymbol{E} = -\operatorname{grad} \operatorname{div} \boldsymbol{E} + \nabla^2\boldsymbol{E} = \nabla^2\boldsymbol{E}$$

Fig. 2.2 Wave vector Q, electric field E and magnetic field B for plane light waves are mutually perpendicular, whereas for sound waves in air the displacement is parallel to Q

since div $E = 0$. A similar derivation applies to B:

$$\Box \Psi = 0 \quad \text{with} \quad \Psi = E, B, A, \text{ and } \phi. \tag{2.16}$$

An important form of solution are the plane waves $\Psi \exp(i\,Qr - i\omega t)$, which wave vector Q and frequency

$$\omega = cQ. \tag{2.17}$$

From this it can be shown that the three vectors E, B and Q are mutually perpendicular: light is a transverse wave and not a longitudinal wave; there are accordingly only two directions of polarisation, not three as in the case of phonons (Fig. 2.2).

2.1.5 Fourier Transformation

The general solution of the wave equation is a superposition of plane waves. For light, a glass prism can physically resolve this superposition into its individual components (colours), just as the ear resolves incident air waves into the individual frequencies (tones). Mathematically, this is called the *Fourier* transformation: a function Ψ of position and/or time can be resolved into exponential functions (waves) of strength $\phi \exp(i\omega t)$, where ϕ depends on the wave vector and the frequency.

We have:

$$\int e^{ixy} dx = 2\pi \delta(y) \tag{2.18}$$

with the integral running from $-\infty$ to $+\infty$. This clearly means that, except for $y = 0$ in the integral, the oscillations cancel each other out. One can derive this formula by multiplying the integrand by a Gauss function $\exp(-x^2/2\sigma^2)$; the integral then gives $\sigma(2\pi)^{1/2}\exp(-y^2\sigma^2/2)$, and for $\sigma \to \infty$ this is the delta function apart from a factor 2π. With this formula it is easily seen that for any arbitrary function of time $\Psi(t)$ we have:

$$\Psi(t) = \frac{1}{\sqrt{2\pi}} \int \Phi(\omega) e^{i\omega t} d\omega \quad \text{with} \quad \Phi(\omega) = \frac{1}{\sqrt{2\pi}} \int \Psi(t) e^{-i\omega t} dt. \tag{2.19}$$

One can also insert a minus sign in the exponential function in the left-hand equation and drop it from the right-hand equation; it is essential, however, that the signs be different. One can also replace the square root of 2π by 2π in one equation, provided one drops this factor altogether in the other equation. The lefthand equation clearly states that the arbitrary function $\Psi(t)$ is a superposition of vibrations $\exp(i\omega t)$ (and hence of cosine and sine waves); these waves each contribute with a strength $\phi(\omega)$ to the total result.

In three dimensions, as functions of position r and wave vector Q, this transformation formula is valid for each of the three components:

$$\Psi(r) = \int \Phi(Q) e^{iQr} d^3 Q / \sqrt{2\pi}^3 \quad \text{with} \quad \Phi(Q) = \int \Psi(r) e^{-iQr} d^3 r / \sqrt{2\pi}^3.$$

If a function is periodic in position or time, then only discrete wave vectors Q and frequencies ω occur. Diffraction of light, x-rays or neutrons in a crystal lattice of solid state physics or a grating in the laboratory are Fourier transformations in space: the light etc. is diffracted only in quite definite directions Q ("Bragg Reflex"). In this way it was first proved by Max von Laue (1879–1960) that solid bodies are periodic arrangements of particles.

2.1.6 Inhomogeneous Wave Equation

Up to this point we have treated the "homogeneous" wave equation $\Box\Psi = 0$: now we consider the "inhomogeneous" case

$$\Box\Psi = -4\pi\rho(r, t),$$

where ρ in the case of the electrical potential (2.14) is the charge density, but in general is an arbitrary function of position and time. The solution is obtained by a sort of Huygens Principle, where Ψ is the superposition of numerous elementary waves spreading outwards from a single point.

Let us consider a bath-tub with a dripping water tap; the water surface permits waves according to $\Box\Psi = 0$, and the falling drops cause such waves, corresponding to the source $-4\pi\rho$. An individual drop has effect only at one definite instant of time t_0 at the point r_0, where it strikes the surface; it therefore corresponds to a delta function $\rho \sim \delta(r - r_0)\delta(t - t_0)$. The effect of one drop is familiar as a circular wave, which starts from the point of origin and then propagates outwards. The effect of all the drops together is the superposition of all these individual circular waves. If the waves are reflected at the side of the bath, then the circular wave is replaced by something more complicated, but the principle of the linear superposition remains the same.

Similarly a delta function $\rho = \delta(r - r_0)\delta(t - t_0)$ in the differential equation $\Box\Psi = -4\pi\rho$ produces a spherical wave about r_0, propagating while $t > t_0$. Boundary conditions can modify these spherical waves; mathematically this spreading wave is then called a Green's function $G(r, t; r_0, t_0)$. Te solution for general density ρ is the superposition of all these Green's functions G produced by the individual delta functions.

If no "sides of the bath-tub" disturb the wave propagation, then the wave Ψ must vanish at infinity. The spherical wave or Green's function as a solution of $\Box G = -4\pi \delta(r - r_0)\delta(t - t_0)$ is then

$$G(r, t; r_0, t_0) = \frac{\delta(t - t_0 - R/c)}{R}, \quad (2.20)$$

where $R = |r - r_0|$. This formula means that a particular distance R is reached by the spherical wave only at a quite definite point in time $t = t_0 + R/c$, if c is its velocity. The factor $1/R$ comes from the energy conservation: since the surface of the sphere increases with R^2, the energy density must decrease as $1/R^2$. This will occur if the amplitude of G decreases as $1/R$, since the energy is proportional to the square of the amplitude, see (2.15). For general densities ρ we solve $\Box \Psi = -4\pi \rho$ by superposition:

$$\Psi(r, t) = \int d^3r_0 \int dt_0 G(r, t; r_0, t_0)\rho(r_0, t_0), \quad (2.21)$$

which can also be directly proved mathematically ($\nabla^2 1/r = -4\pi \delta(r)$). For the electric potential ϕ, ρ is the charge density; for the vector potential A, ρ is replaced by the corresponding component of j/c in order to solve $\Box A = -4\pi j/c$. Since the potentials ϕ and A are to vanish at infinity, (2.20) must hold; after integration over the delta function G in (2.21) we obtain

$$\phi(r, t) = \int d^3r_0 \frac{\rho(r_0, t - R/c)}{R} \quad (2.22)$$

with $R = |r - r_0|$, and the analogous result for A with j/c instead of ρ. In what follows we shall apply these formulae to important special cases.

2.1.7 Applications

(a) Emission of Waves
How do waves propagate outwards when a periodic current $j(r)\exp(-i\omega t)$ flows in a wire ("antenna" of a radio transmitter)? The vector potential A is then given by

$$cA(r)e^{i\omega t} = \int (j(r_0)e^{iQR}/R)d^3r_0$$

with $Q = \omega/c$ and $R = |r - r_0|$, and the scalar potential ϕ by an analogous formula. Everything follows on from here by mathematical approximations, such as, for example, the Taylor expansions

$$R = |r - r_0| = |r| - rr_0/|r| + \dots$$
$$1/R = 1/|r - r_0| = 1/|r| + rr_0/r^3 + \left[3(rr_0)^2/r^5 - r_0^2/r^3\right]/2 + \dots$$

These expansions are valid for distances r large compared with the extent r_0 of the antenna. Substitution in $\exp(iQR)/R$ gives

$$cAe^{i\omega t} = \left(e^{iQr}/r\right)\left[\int j(r_0)d^3r_0 + (r^{-1} - iQ)\int j(r_0)(e_r r_0)d^3r_0 + \dots\right]$$
$$= ce^{i\omega t}\left[A_0 + A_1 + \dots\right]$$

with the "dipole term"

$$rcA_0 = e^{i(Qr - \omega t)}\int j(r_0)d^3r_0.$$

Using the continuity equation $0 = \operatorname{div} j + \rho = \operatorname{div} j - i\omega\rho$ and partial integration, this integral can be rewritten as

$$-\int r_0 \operatorname{div} j(r_0)d^3r_0 = -icQ\int r_0\rho(r_0)d^3r_0 = -icQP,$$

where P is called the electric *dipole moment*. (If for example the charge distribution $\rho(r_0)$ consists only of a positive and a negative charge q at distance d apart, then $|P| = qd$, as is well known for the dipole moment. In an electrical field E the energy of such a dipole moment is equal to $-EP$). The leading term in the above Taylor expansion therefore gives in general

$$A = A_0 + \dots = -iQPe^{i(Qr - \omega t)}/r + \dots$$

The antenna therefore radiates spherical waves, but the amplitude of these waves is proportional to the vector P. The radiation is therefore not equally strong in all directions.

By differentiation we obtain from A the fields B and E:

$$B_0 = \operatorname{curl} A_0 = Q^2(e_r \times P)(1 - 1/iQr)e^{i(Qr - \omega t)}/r \quad \text{and}$$
$$E_0 = (i/Q)\operatorname{curl} B_0$$
$$= \{Q^2(e_r \times P) \times e_r/r + [3e_r(e_r P) - P](r^{-3} - iQr^{-2})\}e^{i(Qr - \omega t)}.$$

These monster expressions become simpler in the "far field", when the distances r from the antenna are very much greater than the wavelength $2\pi/Q$. Then for large

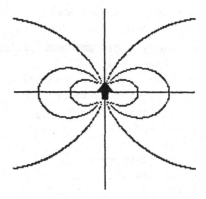

Fig. 2.3 Electric field lines of a dipole pointing upwards. According to (2.23) $|E|$ falls off in a fixed direction as $1/r^3$

Qr we have

$$B_0 = Q^2 e_r \times P e^{i(Qr-\omega t)}/r \quad \text{and} \quad E_0 = B_0 \times e_r.$$

Conversely, for the static limiting case $\omega = 0$ and hence $Qr = 0$ we find that $B_0 = 0$ (charges which are not moving generate no magnetic field). The electric *dipole field*, perhaps already familiar to the reader, is then given by

$$E_0 = [3e_r(e_r P) - P]/r^3. \tag{2.23}$$

Figure 2.3 shows the well known field lines corresponding to (2.23), where at each position these field lines point in the direction of E. (In contrast to the streamlines of steady hydrodynamics the field lines here are not the trajectories of moving electrical charges, except when their inertia can be neglected.) This figure was produced by the simple program DIPOLE in the same way as the Kepler motion: a point (x, y) moves in the direction (E_x, E_y).

The step length dr is multiplied by r^2 in order to economise in calculation time. (The dipole moment is 1; $a \hat{\ } b$ means a^b.) One really needs only those lines whose numbers are integral multiples of 10; the others serve only to call up the graphics routine and need to be adjusted to your computer. Typical starting values are $x = 0.3$, $y = 0.1$; we are concerned here with the principle rather than with precision.

Without derivation, we state that a steady circular current produces a magnetic dipole moment

$$M = \int r_0 \times j(r_0) d^3 r_0/2c$$

exactly analogous to the definition of the electric dipole moment P. The energy of an electric or magnetic dipole in the corresponding field is known to be equal to (–dipole moment · field); this scalar product must of course be negative (energy minimum), when dipole moment and field are parallel. For electric dipoles this follows trivially from energy $= \phi q$ and grad $\phi = -$field, together with the composition of a dipole from two charges q a short distance apart.

```
PROGRAM DIPOLE
    2 hgr: hcolor=7
    4 hplot 130,1 to 130,159
    6 hplot 1,80 to 260,80
   10 input "x,y= "; x,y
   20 dr=0.1
   25 sc=10
   30 r2=x*x+y*y
   40 r5=r2^2.5
   50 ex=(3*x*x-r2)/r5
   60 ey=3*y*x/r5
   70 x=x+ex*dr*r2
   80 y=y+ey*dr*r2
   90 print x,y
   92 hplot 130+sc*x,80+sc*y
   94 hplot 130-sc*x,80+sc*y
   96 hplot 130+sc*x,80-sc*y
   98 hplot 130-sc*x,80-sc*y
  100 if x>0 then goto 30
  110 goto 10
  120 end
```

(b) Multipole Development, ω^4 Law, Moving Point Charges

These were only the leading terms of the Taylor expansion; if one takes still more terms, then one speaks of *multipole* development. If, in (2.22) for the electrical potential $\phi(r)$, we expand the factor $1/R = 1/|r - r_0|$ in powers of r_0, as at the beginning of the preceding section, we obtain quite complicated expressions. These are simple to interpret in the limiting static case, however, when the charges are not moving. The resulting terms, from $1/R \approx 1/r + rr_0/r^3$, are simply

$$\phi = \frac{Q}{r} + \frac{rP}{r^3} \tag{2.24}$$

with the total charge Q and the electric dipole moment P as the space integral over $\rho(r_0)$ and $r_0\rho(r_0)$. The third term gives the so-called quadrupole moment, then comes something similar to Octopussy, each time doubling the power. The arbitrarily complicated charge distribution therefore leads to an expansion which is easy to interpret physically: in the first approximation we place the total charge Q lumped together at the charge centroid as origin ("monopole"), in the second approximation we take in addition an electric dipole P. After that the quadrupole correction can be included as two antiparallel dipoles (i.e. four charges), and so on.

If we turn back to the dynamics of a dipole, we can find its electromagnetic radiation as a surface integral over the Poynting vector (2.15), which is proportional to

Fig. 2.4 Spherical waves
emitted from an oscillator
moving with uniform
velocity

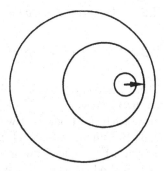

$E \times B$. In the "far field" it was shown in antenna theory that B and E are proportional
to the square of the wave vector Q. For the energy radiated per second, Rayleigh's
law therefore gives

$$\text{radiation} \sim (\text{wavelength})^{-4} \sim \omega^4. \tag{2.25}$$

This explains why the sky is blue and the evening sun is red: blue light has a frequency
about twice as high as red light and is therefore scattered by the air molecules, which
act as little dipoles, very much more strongly than red light. Accordingly, if the sun is
over New York, the blue component of the rays of light falling on Europe is scattered
much more strongly than the red. Europeans therefore receive mainly the red residue
of the evening light, while the New Yorkers see the European blue light in the sky.

If we test these statements about the blue sky frequently, we often see condensation
trails. The noise from these trails appears to come, not from the aircraft, but from an
imaginary source behind the aircraft. This is due to the fact that sound travels very
much more slowly than light. However, even the velocity c of light is finite, so the
electric field of a moving charge q does not appear to come from this, but from an
imaginary point behind the moving charge. When the actual charge is precisely at
the origin, then its potential is not $\phi = q/r$, but $\phi = q/(r - rv/c)$. Now r is the
distance to the previous position from which the light just reached us. A similar
Lienard-Wiechert formula holds for the vector potential: $cA = qv/(r - rv/c)$. One
can see here already one of the foundations of the theory of relativity: no body can
move with a velocity greater than that of light, since not even Einstein could divide by
zero. Figure 2.4 shows, for both sound and light waves, the spherical waves emanating
from a moving source.

2.2   Electrodynamics in Matter

2.2.1   Maxwell's Equations in Matter

There are no new Maxwell equations for matter. Matter consists mostly of pointlike electrons and atomic nuclei, and therefore, for example, the whole of electrostatics is governed by the Coulomb law for point charges. Accordingly, we only need to apply (2.1) 10^{25} times. One could have proceeded in a similar fashion for hydrodynamics, solving 10^{25} coupled Newton's equations of motion. That is not very practical. Just as in hydrodynamics, we solve the problem by averaging: instead of considering each point charge exactly, we average over small regions with many atoms. With these averaged quantities we associate the already introduced charge density ρ and current density j; now we need also the density of the electric and magnetic dipole moments, hence the *polarisation* P and the *magnetisation* M. These approximations lead finally to replacing the electric field E by $D = E + 4\pi P = \varepsilon E$ in some of the Maxwell equations, with a similar formula for the magnetic field. This leads to an electrodynamics for continua in the same sense as the mechanics for continua in Sect. 1.5. (It should be obvious that P and M here do not refer to momentum and torque of the preceding Mechanics chapter.)

If an electric field acts on matter, it can "polarise" the atoms or molecules. For each particle the positive charges (atomic nuclei) are displaced somewhat in one direction, and the negative charges somewhat in the other direction. A small electric dipole moment is thus created in each atom or molecule. In a magnetic field a magnetic dipole moment is similarly created; both types of moment have been defined mathematically in Sect. 2.1.7a. Now let us represent a dipole simply by two electric charges (or a north pole and a south pole) a small distance apart. We define the polarisation P and the magnetisation M as the vector sum of all the dipole moments per unit volume.

If the external electric field changes in the course of time, so will the electric dipoles; i.e. the charges will be displaced and thus cause currents, which have to be taken into account in the Maxwell equations. These so-called polarisation currents j_{pol} have the strength $\partial P / \partial t$. Since an individual dipole moment qd consists of two charges q at a distance d apart, the time derivative $\mathrm{d}(qd)/\mathrm{d}t = q\mathrm{d}(d)/\mathrm{d}t = qv$ is the electric current. We now denote by current density j only the current of conduction electrons "coming out of the wall socket" and therefore have to add to this j the j_{pol}, in order to take account of the displacement of the atoms and molecules by this polarisation current. The Maxwell equation $c\,\mathrm{curl}\,B - \partial E/\partial t = 4\pi j$ without dipole moments thus becomes $c\,\mathrm{curl}\,B - \partial E/\partial t = 4\pi(j + j_{\text{pol}})$ or $4\pi j = c\,\mathrm{curl}\,B - \partial(Ei + 4\pi P)/\partial t = c\,\mathrm{curl}\,B - \partial D/\partial t$ with the *"dielectric displacement"* $D = E + 4\pi P$ (in the electrical units used here).

However, this is still not all. Circular electrical currents cause magnetic dipole moments, and spatial variations in the density M of these magnetic dipole moments lead to additional atomic currents $j_{\text{mag}} = c\,\mathrm{curl}\,M$. Accordingly we have $4\pi(j + j_{\text{mag}}) = c\,\mathrm{curl}\,B - \partial D/\partial t$, or $4\pi j = c\,\mathrm{curl}\,(B - 4\pi M) - \partial D/\partial t = c$

$\operatorname{curl} H - \partial D / \partial t$ with the new field $H = B - 4\pi M$ analogously to the definition of D.

As the third and last atomic contribution we take into account the fact that local so-called polarisation charges ρ_{pol} can form, in addition to the charge density ρ of free electrons or ions. Of course, if many dipoles get positive charges attached to their heads, then a positive excess charge is caused; the compensating negative charges of the dipoles occur elsewhere. If vector arrows put their heads together, this head loading is described by the divergence of the vector field; hence we get $\rho_{\text{pol}} = -\operatorname{div} P$. Substitution in the original Maxwell equation $\operatorname{div} E = 4\pi\rho$ now gives $\operatorname{div} E = 4\pi\rho - 4\pi \operatorname{div} P$ or $\operatorname{div} D = 4\pi\rho$. Fortunately, now as before, there are no magnetic monopoles, $\operatorname{div} B = 0$, and also $0 = c \operatorname{curl} E + \partial B / \partial t$ does not change, in the absence of magnetic monopole currents.

We have now clarified all the Maxwell equations in matter:

$$D = E + 4\pi P; \quad \operatorname{div} D = 4\pi\rho, \quad c \operatorname{curl} H - \frac{\partial D}{\partial t} = 4\pi j \quad (2.26)$$

$$B = H + 4\pi M; \quad \operatorname{div} B = 0, \quad c \operatorname{curl} E + \partial B / \partial t = 0. \quad (2.27)$$

Anybody who cannot remember where E and D go, will be pleased to learn that the energy density is quite symmetrical, namely $(ED + HB)/8\pi$.

2.2.2 Properties of Matter

We now indeed have the Maxwell equations, but we do not know how great are the polarisation P and the magnetisation M. Both depend on the material under investigation. In statistical physics we shall learn to calculate M (and similarly P); here we shall be content to take M and P from experiment.

There are materials in which M (or P) are non-zero even in the absence of an external field. In the elements iron, cobalt and nickel at room temperature (and in gadolinium when it is cold) there is *spontaneous magnetisation* of this sort without a magnetic field; they are called *ferromagnets*. Similarly there are also *ferroelectric* materials, such as potassium dihydrogen phosphate, which shows spontaneous polarisation even without an external electric field. If ferromagnets are heated above their *Curie temperature* the spontaneous magnetisation vanishes (*paramagnetism*). Just below this Curie temperature the spontaneous magnetisation varies as $(T_c - T)^\beta$, with the critical exponent β close to 1/3 for ferromagnets and 1/2 for ferroelectrics.

If there is no spontaneous magnetisation (spontaneous polarisation), then for weak external fields the magnetisation M (or polarisation P) is proportional to the field. In the electric field an electric current j also flows:

$$j = \sigma E, \quad P = \chi_{\text{el}} E, \quad M = \chi_{\text{mag}} H \quad (2.28)$$

with the conductivity σ and the electric or magnetic *susceptibility* χ_{el} or χ_{mag}. From the above definitions for D and H we then have

$$D = (1 + 4\pi \chi_{el})E = \varepsilon E, \quad B = (1 + 4\pi \chi_{mag})H = \mu H \qquad (2.29)$$

with the *dielectric constant* ε and *the permeability* μ. It is customary to work with ε for electric phenomena and with χ for magnetic phenomena. (In principle, all the proportionality factors σ, χ, ε and μ introduced here are tensors.) All these laws of "linear response" hold only for sufficiently small external fields; the nearer to the Curie point, the smaller must the fields be, until at $T = T_c$ itself this linear approximation breaks down altogether and χ becomes infinite.

One can also calculate the spontaneous magnetisation on the computer (program ISING). We set a "spin" IS (atomic magnetic dipole moment) at each lattice position of a square lattice; $IS = 1$ or -1 according to whether the spin is up or down. Neighbouring spins want to be parallel in this *"Ising"*-ferromagnet of 1925. The energy will thus be conserved in a reversal of spin, if as many of the neighbouring spins are up as are down, i.e. if the sum over the four neighbouring IS is zero. In this sense the program reverses a spin $[IS(i) = -IS(i)]$ if, and only if, the sum over the neighbouring spins vanishes.

The spins of an $L \times L$ lattice are stored in a one-dimensional array $IS(i)$, with $i = L + 1$ for the first and $i = L^2 + L$ for the last spin. The left-hand neighbour then has the index $i - 1$, the right-hand one the index $i + 1$, the one above the index $i - L$ and the one below the index $i + L$, if one runs through the lattice as with a typewriter. There are, in addition, two buffer lines at the upper ($1 \leq i \leq L$) and lower ($L^2 + L + 1 \leq i \leq L^2 + 2L$) boundaries, so that all lattice points have neighbours ("helical boundary conditions"). Initially the spins are randomly oriented: $IS = 1$ with probability p, otherwise $IS = -1$. This is achieved by comparing a random number RND, which lies anywhere between 0 and 1, with the probability p, so that $RND < p$ with probability p. We leave it to the computer how to play dice in order to calculate RND (*"Monte Carlo* simulation").

PROGRAM ISING

```
10  dim is (1680)
20  L=40
30  p=0.2
40  L1=L+1
50  Lp=L*L+L
60  Lm=Lp+L
70  for i=1 to Lm
80  is(i)=-1
90  if rnd(i)<p then is(i)=1
100 next i
110 for it=1 to 100
120 m=0
130 for i=L1 to Lp
140 if is(i-1)+is(i+1)+is(i-L)+is(i+L)=0 then is(i)=-is(i)
150 m=m+is(i)
160 next i
170 print it,m
180 next it
190 end
```

For $p = 0$ or $p = 1$ all the spins are always parallel, which corresponds to quite low temperatures. For $p = 1/2$ the spins are randomly oriented, which corresponds to very high temperatures (magnetisation $M = 0$ apart from fluctuations). At $p = 0.08$ the Curie point is reached: for $p > 0.08$ the magnetisation goes slowly to zero, for $p < 0.08$ it retains a finite value, equal to the spontaneous magnetisation.[1]

2.2.3 Wave Equation in Matter

One of the earliest scientific experiments on electrodynamics in matter was the treatment of light refraction into glass or water: the light path is no longer rectilinear, but such that the light travels as quickly as possible from the starting point to the endpoint ("Fermat's Principle"). What is the velocity of light c_{medium} in a medium described by (2.28)? We consider an isotropic insulator such as glass: $j = 0 = \rho$, so that c curl $E = -\partial B / \partial t$ and c curl $H = \partial D / \partial t$. If we differentiate the last equation with respect to time and substitute the first [as previously in (2.16)], we obtain for $c^{-2} \partial^2 D / \partial t^2$:

$$
\begin{aligned}
(1/c)\partial(\text{curl } H)/\partial t = (1/c)\text{curl}(\partial H/\partial t) &= (1/c\mu)\text{curl}(\partial B/\partial t) \\
&= -(1/\mu)\text{curl curl } E = -(1/\mu)(\text{grad div } E - \nabla^2 E) \\
&= -(1/\mu)(\text{grad div } D/\varepsilon - \nabla^2 D/\varepsilon) = 1/(\mu\varepsilon)\nabla^2 D
\end{aligned}
$$

---

[1]H.-J. Herrmann, J. Stat. Phys. 45, 145 (1986).

Fig. 2.5 Integration path
for calculation of normal and
tangential components of D
and E at a surface

and hence

$$c_{\text{medium}} = \frac{c}{\sqrt{\mu\varepsilon}}. \tag{2.30}$$

The refractive index n is given by the ratio of the velocities of light

$$n = \sqrt{\mu\varepsilon}, \tag{2.31}$$

and determines the ratio of the sines of the incident angle and the refracted angle in the refraction law. (In water $\mu \approx 1$ and $\varepsilon \approx 81$; nevertheless the refractive index is not 9, but about 4/3. This is related to the fact that ε is frequency dependent: light takes the low value for high frequencies, whereas $\varepsilon \approx 81$ corresponds to zero frequency. Because of this frequency dependence of ε the different colours are strongly separated by a glass prism, i.e. the light is "resolved".)

2.2.4 Electrostatics at Surfaces

In an electrical conductor the electric field is always zero at equilibrium, since otherwise currents would flow. In an insulator, on the other hand, there may be fields even in equilibrium; the electric current is still zero. Accordingly we have: $H = B = j = \text{curl}\, E = 0$ and $\text{div}\, D = 4\pi\rho$. The laws of Gauss and Stokes then state that

$$\oint E dl = 0 \quad \text{and} \quad \oint D d^2 S = 4\pi Q$$

with the charge Q. For the integration regions we take those sketched in Fig. 2.5: For the Stokes law we take a long flat rectangular loop, and for the Gauss law we take two planes lying close to each other and enclosing the surface. The figure shows the loop; for the planes one has to imagine the picture continued in space in front of and behind the plane of the paper, with uniform width L. The two narrow portions normal to the boundary area make a negligible contribution to the integrals. The calculation is valid in general, but one can visualise the first medium as air (vacuum, $\varepsilon_1 = 1$) and the second as glass ($\varepsilon_2 > 1$).

Then it follows from the Stokes loop of length L that $E_2^{\text{tang}} L - E_1^{\text{tang}} L = 0$ for the tangential component of E parallel to the surface. From the Gauss integral one obtains $D_2^{\text{norm}} L^2 - D_1^{\text{norm}} L^2 = 4\pi Q$, where Q is the electric charge between the two

Fig. 2.6 Induction with a
point charge in front of a
metal plate. The shaded
lower half is metal, and in
reality contains no field lines
and no negative image
charge

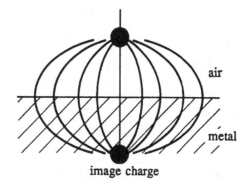

image charge

integration planes of area L^2. With the surface charge density $\sigma = Q/L^2$ (charge
per square centimetre) we therefore have:

$$E_1^{\text{tang}} = E_2^{\text{tang}} \quad \text{and} \quad D_1^{\text{norm}} = D_2^{\text{norm}} - 4\pi\sigma,$$

or, even more simply (without surface charges):

> The tangential component of E and the normal component of D
> are continuous across the surface.   (2.32)

Such *surface charges* can be present on both glass and on metal surfaces. On the
glass surface electrically charged particles can accumulate from the air in the course
of time; the metal surface, on the other hand, can receive electrons from the interior
by *induction*. The number of electrons accumulating on the metal surface is just that
number required to ensure that in equilibrium there is no field parallel to the surface.
In the interior of the metal E is in any case zero. Accordingly, if the index 2 describes
the metal, $E_2 = 0$, $E_1^{\text{tang}} = 0$ and $E_1^{\text{norm}} = -4\pi\sigma/\varepsilon_1$. The field E_1 can be caused,
for example, by a positive point charge in the vacuum in front of the metal plate, as
sketched in Fig. 2.6 (program similar to that for Fig. 2.5).

By means of the Gauss law it can be shown that the induced total negative charge
is exactly as great as the positive charge. This is valid even for twisted surfaces. In the
special case of the plane surface of Fig. 2.6 the field in the vacuum is the same as if the
metal with its surface charge were not there but were replaced by an equal negative
point charge at an equal distance below the surface. This "image charge" shows that
a point charge in front of a metal surface causes a dipole field in the vacuum. The
trick with the image charge is also very useful elsewhere in electrostatics.

Table 2.1 Electrostatic fields in the medium and its cavities

	E	P	D
Medium	$4\pi\sigma/\varepsilon$	$(\varepsilon-1)\sigma/\varepsilon$	$4\pi\sigma$
Longitudinal slot	$4\pi\sigma/\varepsilon$	0	$4\pi\sigma/\varepsilon$
Transverse slot	$4\pi\sigma$	0	$4\pi\sigma$
Vacuum	$4\pi\sigma$	0	$4\pi\sigma$

In magnetostatics there are no magnetic monopoles and therefore no surface charges; accordingly the tangential component of H is always continuous across the surface, as is the normal component of B. If iron (large spontaneous magnetisation, $B = H + 4\pi M \approx 4\pi M$) borders on air ($B = H$ since $M = 0$), then the vector of magnetisation must be almost parallel to the surface, since B^{norm} is continuous and hence is small. The magnetic field lines of a transformer are therefore "imprisoned" by the iron core threading the two coils, which are thus linked magnetically, though insulated from each other electrically. Accordingly, at each cycle of the alternating current, c curl $E = -\partial B/\partial t$ is the same for each loop of wire in each coil, and the voltage is transformed in the ratio of the number of loops in the two coils. A transformer might appear more beautiful with oak instead of iron, but it would not work so well.

Finally, what is a *condenser*? Suppose we place a glass plate (thickness d) with dielectric constant ε between two square copper plates (side L), having a voltage U between them, applied from an external source. Equal and opposite surface charge densities of magnitude $\sigma = Q/L^2$ are thus created on the copper plates. Copper is metallic and not ferroelectric, so $E = D = 0$ in the copper. On the interfaces between the glass and the copper we therefore have $D^{\mathrm{norm}} = 4\pi\sigma$ in the glass; the tangential component is zero. Accordingly, in the glass we have $4\pi\sigma = D = \varepsilon E = \varepsilon U/d$. The charge density is therefore given by

$$\sigma = \varepsilon U/(4\pi d).$$

The larger ε is and the smaller the distance d between the plates, the larger is the capacitance $Q/U = \varepsilon L^2/(4\pi d)$.

By similar methods Table 2.1 treats the various geometries of Fig. 2.7, with in each case the upward pointing vectors D, E and $P = (D - E)/4\pi = (\varepsilon - 1)E/4\pi$.

The field acting on an atom in a material is neither D nor E. Let n be the number of such atoms per cubic centimetre, and α the polarisability of an isolated atom (i.e. the ratio of dipole moment to applied field). Then people more intelligent than this author have derived the approximative *Clausius-Mossotti* formula:

$$\varepsilon = (1 + n\alpha 8\pi/3)/(1 - n\alpha 4\pi/3).$$

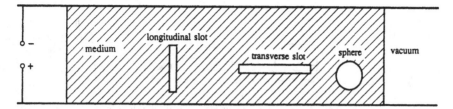

Fig. 2.7 Various shapes of cavity in a dielectric such as glass

This provides a simple model for ferroelectricity: if the polarisability α of the individual atom is so large that $n\alpha$ comes close to $3/4\pi$, then ε becomes very large. A small external field first polarises the atoms, whose resulting atomic dipolar moments strengthen the field and so increase the polarisation still further, and so on. If $n\alpha = 3/4\pi$, this feedback leads to polarisation catastrophe: ε becomes infinite, and the Curie point is reached. In this picture, of course, we have omitted the effect of the temperature, which is only taken into account in Statistical Mechanics (mean field theory).

2.3 Theory of Relativity

Michelson and Morley established experimentally in 1887 that the velocity of light c is the same on the earth in all directions and is not affected by the motion of the earth around the sun etc. Thus, whereas sound waves consist of motions in an elastic medium, relative to which they have a fixed velocity, it appears that for electromagnetic waves there is no such medium ("ether"). (We are not now considering light propagation in matter, Sect. 2.2.3). Instead, we have the principle of relativity:

> **The laws of physics are the same in all inertial systems.**

Since Maxwell's equations *in vacuo* have been repeatedly verified, the velocity of light c *in vacuo* derived from them must be the same in all inertial systems. This does not mean that "everything is relative", or that there is no absolute velocity in the universe. From the presumed primeval "big bang" of the universe more than 10^{10} years ago, electromagnetic waves still survive, and even exert "light pressure" (Sect. 2.1.3). It has been shown experimentally that this radiation pressure of the so-called 3-Kelvin background radiation is variable in strength from different directions of the universe: the earth moves at several hundred kilometres per second relative to the "big bang" radiation. The Lorentz force $F/q = E + v \times B/c$ on the other hand is a general law of nature and therefore valid in all inertial systems. What can the velocity v here mean then: velocity from where? Albert Einstein solved this problem

Fig. 2.8   Example of a stationary (x, t) and a moving (x', t') inertial system. Relativistic corrections are negligible for the Paris-Cologne express

in 1905 in his work on the electrodynamics of moving bodies, generally known today as the Special Theory of Relativity.

2.3.1   Lorentz Transformation

(a) Derivation
Before we consider electromagnetic fields relativistically, let us first discuss the fundamentals: the transformation of position and time. For simplicity we assume a one-dimensional motion. If in an inertial system an event occurs at time t and position x, what are the coordinates x' and t' in another inertial system, moving with velocity v relative to the first system? (Let the origin of coordinates be initially the same.) Classical mechanics gives the simple answer in the Galileo transformation (Fig. 2.8):

$$x' = x - vt, \quad t' = t.$$

Light waves emitted from a body moving with velocity v now have the velocity $c \pm v$, since in the Galileo transformation the velocities are simply additive.

The Michelson-Morley experiment, and many other results in the past hundred years, however, show that the velocities are not simply additive; instead, the following postulates hold:

1. The velocity of light c *in vacuo* is constant.
2. In all inertial systems force-free motions are unaccelerated (uniform and in a straight line).
3. No reference system is preferred to another in the transformation of position and time.

Mathematically, postulate 1 means that: $x' = ct'$ if $x = ct$; hence no new c'! From postulate 2 it follows that position and time are transformed linearly:

$$x' = ax - bt; \quad t' = Ax - Bt.$$

Postulate 3 means that if we know the transformation coefficients a, b, A and B as functions of the velocity v, then the reverse transformation $x = x(x', t')$ and $t = t(x', t')$ ("inverse matrix"), follows by replacing v by $-v$. Now we can determine these coefficients purely by mathematics:

The origin of the one system moves with velocity v relative to the other. If we have for the origin $x = vt$, then $x' = 0 = avt - bt$, so $b = av$.

In general it therefore follows that $x' = a(x - vt)$ and according to postulate 3: $x = a(x' + vt')$. Postulate 1 (if $x = ct$ then $x' = ct'$) now takes the form: if $ct = at'(c + v)$, then $ct' = at(c - v)$, and hence

$$ct' = a\left[at'(c + v)/c\right](c - v) \quad \text{or} \quad a = 1/\sqrt{1 - v^2/c^2} = \gamma.$$

This square root expression will be repeatedly encountered in the theory of relativity; it is usually denoted by γ instead of a. It follows that $b = v\gamma$ and $x' = \gamma(x - vt)$. From postulate 3 we now find that

$$x = \gamma(x' + vt') = \gamma(\gamma x - \gamma vt + vt') \quad \text{or}$$
$$t' = \gamma t + x(1 - \gamma^2)/\gamma v = \gamma(t - vx/c^2).$$

This determines the other two coefficients A and B and we can collect together this "*Lorentz transformation*":

$$x' = \gamma(x - vt), \quad t' = \gamma\left(t - \frac{vx}{c^2}\right) \quad \text{with} \quad \gamma = \frac{1}{[1 - v^2/c^2]^{1/2}}. \tag{2.33}$$

In three dimensions we also have $y' = y$ and $z' = z$.

(b) Deductions
Division into Classes. Because of the constancy of the velocity of light we have $x^2 - c^2 t^2 = x'^2 - c^2 t'^2$ (also provable directly). Similarly in three dimensions the quantity $r^2 - c^2 t^2$ is Lorentz invariant, i.e. it does not change in a Lorentz transformation, just as the scalar product r^2 does not change with rotation of the axes. For two events separated by a spatial distance r and a time interval t we can therefore establish a Lorentz invariant three-way classification:

$$\underbrace{light - like}_{r^2-c^2t^2=0} \quad \underbrace{space - like}_{r^2-c^2t^2>0} \quad \underbrace{time - like}_{r^2-c^2t^2<0} .$$

When a cause propagates more slowly than with the velocity of light (and this is true of all known methods of conveying energy or information), then the relation between cause and effect is time-like, when seen from any arbitrary inertial system: causality is Lorentz invariant. If there were particles which travelled faster than light (e.g., "tachyons" with imaginary mass), then the causality would be problematic, and many causes would arise only after their effects.

Travellers Stay Young. An accurate clock lies at the origin of the x' system, which moves with velocity v relative to the x system of the observer. How fast does it tick? Because $x' = 0$ and $x = vt$ we have $t' = \gamma(t - v^2t/c^2) = t(1 - v^2/c^2)^{1/2}$, whence $t' < t$. [Alternatively: $t = \gamma(t' + vx'/c^2)$ in general and here $x' = 0$, so $t = \gamma t'$.] Accordingly, for the clock to show a time t' of one hour, the observer would have to wait $t = 10$ h, if $v = 0.995c$. This was demonstrated for myons (earlier falsely called μ-mesons) in cosmic radiation. With a lifetime of about 10^{-6} s they should certainly not reach the surface of the earth, when they are created at a height of a few kilometres by the collision of high energy particles with air molecules. Nevertheless many of these myons reach us, the extension of their lifetime by the factor γ allowing them to follow a much longer path. In modern particle accelerators this extension of the lifetime has been determined much more precisely. Originally one could scarcely believe that, of two twins, that one which has made a journey through outer space should finally be much younger than that one which had remained on earth. One spoke of the twins paradox.

Jogging Makes One Slim. A metre rule is at rest in the x' system, which moves with a velocity v relative to the x system of the observer. How long is the rule for the observer? We measure the length with the left-hand end (index 1) at time $t_2 = t_1 = t_1' = 0$ at the common origin of the two inertial systems. Accordingly $1 \text{ m} = x_2' = \gamma(x_2 - vt_2) = \gamma x_2$ or $x_2 < 1$ m. The observer, who measures the left-hand and right-hand ends at the *for him* same times t_1 and t_2, therefore observes a shortening of the length of the moving rule. So can you park your Cadillac in a small garage!??

Time dilatation and length contraction can be linked together: lengths and times change by the factor

$$\gamma = \frac{1}{\sqrt{1 - v^2/c^2}}. \tag{2.34}$$

Addition of Velocities. A train travels from Cologne to Paris with velocity v, and a passenger moves westwards through the dining-car with velocity u. What velocity x/t does the passenger have, relative to the ground? Let the x system be the ground, and the x' system the train. Then $x' = ut'$ for the passenger. The Lorentz transformation $x = \gamma(x' + vt')$ implies that for him: $x = \gamma(u + v)t'$. Similarly we have

$$t = \gamma(t' + vx'/c^2) = \gamma(1 + uv/c^2)t'.$$

The compound velocity is therefore

$$\frac{x}{t} = \frac{u + v}{1 + uv/c^2}. \tag{2.35}$$

the classical addition $x/t = u + v$ being valid only for velocities far below the velocity of light. In particular, (2.35) shows that the total velocity is again $x/t = c$ if either of the two velocities u or v (or both) is equal to the velocity of light. The light of a star moving towards us therefore travels towards us with exactly the same velocity, namely c, as if it were moving away from us. Using double stars, astronomers have verified this with great precision.

Initially controversial, this Special Theory of Relativity is today generally accepted and is the staple diet of high energy physics. The General Theory of Relativity, developed a decade later, (space curvature by mass, leading to the black hole and gravitational waves) is much more complicated, less precisely confirmed, and still leaving alternatives. Presumably, however, Einstein is correct here too.

2.3.2 Relativistic Electrodynamics

If a bar magnet is introduced into a coil, the area integral over the magnetic field B is thereby changed, and since $\partial B/\partial t = -c\,\mathrm{curl}\,E$ this induces an electric field in the loop. If the bar magnet is held stationary and the coil is moved towards it with velocity v, then no E-field is created, but instead, the Lorentz force $qv \times B/c$ moves the electrons in the coil. Viewed relativistically, however, both processes are equivalent; the fields E and B are therefore only different manifestations of the same underlying field. We shall become familiar with it as the 6 components of a four-dimensional antisymmetric field tensor.

The fourth dimension is, of course, time, and with the division into light-like, space-like and time-like classes we have already used the four-dimensional scalar product $r^2 - c^2t^2$. We therefore define an imaginary length $x_4 = ict$ as the fourth component and apply the usual definition of the scalar product

$$xy = \sum_\mu x_\mu y_\mu \quad \text{with} \quad \mu = 1, 2, 3, 4$$

and call this four-dimensional space-time continuum after Minkowski, who suggested it in 1908 at a conference in Cologne.

There is nothing imaginary about time in relativity; the factor i is just mathematically convenient to produce the minus sign in $r^2 - c^2t^2$. Somewhat similarly we used complex numbers for oscillations with friction in Sect. 1.1.4 without claiming anything imaginary about this motion. To prepare for General Relativity with big masses curving the space, one may use a metric tensor $g_{\mu\nu}$ in Minkowski space which is $(1, 1, 1, -1)$ along the main diagonal and zero elsewhere as long as we deal with Special Relativity. Then the space-time vector is (x, y, z, ct) without a factor i and the scalar product of two such vectors becomes

$$xy = \sum_{\mu} \sum_{\nu} x_{\mu} g_{\mu\nu} y_{\nu}.$$

In General Relativity the off-diagonal elements can become non-zero. But now we return to our primitive ict mathematics.

The wave operator $\Box = \nabla^2 - c^{-2}\partial^2/\partial t^2$ is now simply $\sum_{\mu} \partial^2/\partial x_{\mu}^2$. Scalar products are called so because they are scalar, and therefore invariant in a rotation of the coordinate axes (here a Lorentz transformation). The vector of the x_{μ} in a Lorentz transformation is multiplied by the matrix

$$L = \begin{pmatrix} \gamma & 0 & 0 & i\gamma v/c \\ 0 & 1 & 0 & 0 \\ 0 & 0 & 1 & 0 \\ -i\gamma v/c & 0 & 0 & \gamma \end{pmatrix}$$

to give the transformed space-time vector x_{μ}'.

As well as the 4-vector $(r, ict) = (x_1, x_2, x_3, x_4)$ for position and time there are also other 4-vectors, which are transform like the position-time vector. These include the current density tetrad $(j, c\rho)$ and the 4-potential $(A, i\phi)$. The wave equation $c\Box A_{\mu} = -4\pi j_{\mu}$ is valid now for all four components, and the continuity equation $\mathrm{div}\, j + \partial\rho/\partial t = 0$ has the form of a divergence tetrad: $\sum_{\mu} \partial j_{\mu}/\partial x_{\mu} = 0$. The Lorentz gauge reads $\sum_{\mu} \partial A_{\mu}/\partial x_{\mu} = 0$. These simplifications in the notation indicate that we are on the right road.

We define a 4×4 antisymmetric field tensor $f_{\mu\nu}$ by

$$f_{\mu\nu} = \partial A_{\mu}/\partial x_{\nu} - \partial A_{\nu}/\partial x_{\mu}$$

with $\mu, \nu = 1, 2, 3, 4$. Since $f_{\mu\nu} = -f_{\nu\mu}$ this tensor has only six independent matrix elements:

$$f_{\mu\nu} = \begin{pmatrix} 0 & B_z & -B_y & -iE_x \\ -B_z & 0 & B_x & -iE_y \\ B_y & -B_x & 0 & -iE_z \\ iE_x & iE_y & iE_z & 0 \end{pmatrix}. \tag{2.36}$$

As a properly defined tensor, the field tensor f must behave according to the rules for transformation of matrices in a Lorentz transformation and link together 4-vectors before and after. Therefore E fields transform into B fields and vice versa: E and B are different forms of the same field. From the standpoint of relativity theory, therefore, it is not sensible to measure E and B in different units; many relativists even set $c = 1$ and measure position and time in the same units. Thus no volt and ampere here.

There is therefore no contradiction between the magnet moving into the coil, and the coil which moves round the magnet without an E field: the division of the electromagnetic effects between E and B depends on the system of reference. Without motion there is only the E field, and so the B field is the relativistic correction to the electric field. In a three-dimensional coordinate transformation, E rotates like a three-dimensional vector, since E represents a line or column of the field tensor f. On the other hand B is not a genuine (polar) vector, but an antisymmetric 3×3 matrix, as we see in (2.36).

In order to show that the quantities E and B defined in (2.36) really represent the electric and magnetic fields, we have still to derive Maxwell's equations from them. This can be done using the calculation rules

$$\sum_{\nu} \partial f_{\mu\nu}/\partial x_\nu = (4\pi/c)\, j_\mu \quad \text{and} \quad \partial f_{\mu\nu}/\partial x_\lambda + \partial f_{\nu\lambda}/\partial x_\mu + \partial f_{\lambda\mu}/\partial x_\nu = 0.$$

The 4-force density, whose first three components are ρE for stationary charges, is given in general in its μ-th component by $\Sigma_\nu\, f_{\mu\nu}\, j_\nu/c$, which leads to the Lorentz force proportional to $E + v \times B/c$ and justifies the name field tensor for $f_{\mu\nu}$. In this sense therefore the Maxwell equations are the relativistic generalisations of the Coulomb law. Relativity theory does not lead to corrections for the Maxwell equations, but just makes them clearer. Transformers work relativistically!

2.3.3 Energy, Mass and Momentum

Now that we have become familiar with the advantage of the tetrad notation for E and B, we shall turn back once again to mechanics, where we have already come across $(x_\mu) = (r, ict)$. (The round brackets in (x_μ) and other 4-vectors will distinguish the 4-vector as an entity from its four components x_μ.) Other 4-vectors must be transformed in a Lorentz transformation exactly as these position-time vectors are. A 4-scalar such as $\Sigma_\mu x_\mu^2 = r^2 - c^2 t^2$, on the other hand, does not change at all in a Lorentz transformation. The product of a vector with a scalar is again a vector.

Since the time t is not a scalar, (dx_μ/dt) is not a 4-vector. But $r^2 = t^2 - r^2/c^2$ is a scalar, and the same is true of the differential

$$dr = \sqrt{dt^2 - dr^2/c^2} = \sqrt{-\sum dx_\mu^2}/c = dt/\gamma,$$

the "eigentime". For the 4-velocity and the 4-acceleration, therefore, we differentiate with respect to τ and not t:

$$4\text{-velocity } (v_\mu) = (dx_\mu/d\tau) = \gamma(\boldsymbol{v}, ic)$$
$$4\text{-acceleration}(\alpha_\mu) = (dv_\mu/d\tau) = \gamma d(\gamma(\boldsymbol{v}, ic))/dt.$$

Here the three-dimensional velocity $\boldsymbol{v}$ is, as usual, the derivative of the position with respect to t, not τ. The 4-force is now

$$(F_\mu) = (a_\mu)m_0 = m_0\gamma d(\gamma(\boldsymbol{v}, ic))/dt = (\boldsymbol{F}, icm_0d\gamma/dt)\gamma$$

with m_0 the rest mass. If one defines the velocity-dependent mass

$$m = \gamma m_0 = m_0/\sqrt{1 - v^2/c^2},$$

then the fourth component of the 4-force is $ic\gamma\, dm(v)/dt$, and the threedimensional force $\boldsymbol{F}$

$$\boldsymbol{F} = \frac{d(m\boldsymbol{v})}{dt} \quad \text{with} \quad m = m(v) = \gamma m_0. \tag{2.37}$$

The 4-momentum is defined as

$$(p_\mu) = (v_\mu)m_0 = m_0\gamma(\boldsymbol{v}, ic) = (m\boldsymbol{v}, icm) = (\boldsymbol{p}, icm)$$

with the three-dimensional momentum $\boldsymbol{p} = m\boldsymbol{v} = \gamma m_0\boldsymbol{v}$. The mass therefore becomes the greater, the greater the velocity v is, and becomes infinite when $v = c$. It therefore takes more and more effort to accelerate a particle to velocities near c, which is noticeable in the budgets for Fermilab (Chicago), Large Hadron Collider (Geneva) and other accelerators. Newton's law of motion is valid now only in the form $\boldsymbol{F} = d\boldsymbol{p}/dt$ and no longer in the form $\boldsymbol{F} = md\boldsymbol{v}/dt$.

The scalar product $\Sigma_\mu p_\mu^2 = m_0^2\gamma^2(v^2 - c^2) = -m_0^2c^2$ is constant The same is true of $\Sigma_\mu v_\mu^2 = -c^2$. Accordingly we also have

$$0 = d\left(\sum v_\mu^2\right)/d\tau = \sum 2v_\mu dv_\mu/d\tau = \sum v_\mu a_\mu$$
$$= (2/m_0)\sum v_\mu F_\mu = (2\gamma^2/m_0)\left(\boldsymbol{v}\boldsymbol{F} - c^2dm/dt\right),$$

or

$$\text{power } \boldsymbol{v}\boldsymbol{F} = d(mc^2)/dt.$$

Since power = change of energy per unit time, we obtain for the energy the most renowned formula of the theory of relativity:

$$E = mc^2, \tag{2.38}$$

derived here by gentler methods than the hydrogen bomb. The 4-momentum is then recognised as the combination of momentum and energy: $(p_\mu) = (\boldsymbol{p}, iE/c)$ and the scalar product $\Sigma p_\mu^2 = -m_0^2$ is $p^2 - E^2/c^2$:

$$E = \sqrt{(m_0 c^2)^2 + p^2 c^2}. \tag{2.39}$$

For high velocities the momentum term dominates here: $E = pc$ as for photons (light quanta), neutrinos, and electrons in the GeV region. For low velocities we expand the square root as $(a^2 + \varepsilon)^{1/2} = a + \varepsilon/2a$:

$$E = m_0 c^2 + p^2/2m_0 + ... = E_0 + E_{\text{kin}}.$$

The formula of classical mechanics, kinetic energy $= p^2/2m$, is therefore not false, but only the limiting case of the theory of relativity for small velocities. (For a critique of this presentation see L.B. Okun, Physics Today, June 1989, page 31, and May 1990, page 13.) Sometimes in theoretical physics, the truths of one century are the half-truths of the next century and only special limits.

Questions

Section 2.1

1. What form do the Coulomb force and the Coulomb potential take for a test charge e, if a charge q is located at the origin?
2. State Maxwell's equations for the steady case.
3. What is the force acting on a charge q moving in the fields $\boldsymbol{E}$ and $\boldsymbol{B}$?
4. What are the relationships between energy density, energy flux density and electromagnetic fields?
5. What is the significance of the Fourier transformation?
6. What is the form of the scalar potential, when the charge density ρ is known as a function of position and time? Interpretation?
7. What is the significance of the Green's function for inhomogeneous linear differential equations such as $\square \phi = -4\pi\rho$?
8. How do you make an electric dipole?
9. What is the force on a magnetic dipole in a homogeneous magnetic field?

Section 2.2

10. What is the form of the potential for charges moving with the velocity of light c?
11. Do Maxwell's equations change inside matter?

Section 2.3

12. How could you conveniently produce velocities $>c$?
13. By how much would your life be lengthened if you danced all night?
14. How are Maxwell's equations changed by relativity?
15. What is four-momentum, and how does it transform?

Problems

Section 2.1

1. What is the electric potential of a charged sphere, if the charge density depends only on the distance from the centre of the sphere, and not on the direction (onion skin)? What follows analogously for the gravitational acceleration in mineshafts and aeroplanes?
2. How does a transformer work? For simplicity consider two interwoven coils with the same radius but different numbers of windings in each coil.
3. What are the Fourier transforms of $\exp(-|t|/\tau)$ in one dimension, and of $\exp(-r/\xi)/r$ in three dimensions?
4. When do earthworms take their Christmas holiday, i.e., how does a temporal sine wave of the temperature propagate into the interior of the earth according to the diffusion equation?
5. An electron orbits at a distance r around a proton: what is the ratio of its magnetic dipole moment to its angular momentum; with what (in order of magnitude) angular velocity ω does it orbit when $r = 1\,\text{Å}$?
6. How do you make experimentally the Green's function of the diffusion equation, analogous to the already familiar Green's function of the wave equation?

Section 2.2

7. What is the qualitative form of the electrostatic potential if a dipole is suspended parallel to, and close to, a metal surface?

Section 2.3

8. Calculate explicitly how $r^2 - c^2 t^2$ changes in a Lorentz transformation.
9. How can you measure time dilatation without accelerations occurring?
10. Show that curl $E = 0$ for the field E defined by the relativistic field tensor, in the static case.

Chapter 3
Quantum Mechanics

Abstract The energy balance $E = h\nu$ of the quantum-mechanical photo effect looks so simple but we have to go to the unsteady perturbation theory at the end in order to understand it.

The theory of relativity has already shaken the foundations of our understanding of the natural world: two events, which for one observer occur at the same time, for another observer are not simultaneous. Quantum mechanics goes still further: a particle is no longer in a definite place at a definite point in time; instead, we have Heisenberg's Uncertainty Principle. In the theory of relativity the difficulties come from the fact that the velocity is finite and therefore, according to the observer, two mutually remote events are seen as simultaneous or not simultaneous. In quantum mechanics the uncertainty comes from the fact that one has only a certain definite probability for the occurrence of a particle or other event. Both in relativity and in quantum mechanics it is therefore meaningful only to use such quantities and concepts as one can in principle also measure.

3.1 Basic Concepts

3.1.1 Introduction

Do we actually need quantum mechanics, once we escape from the discipline of the university? Up to now, we have been unable to resolve a number of important questions:

Atomic Structure. Why do the electrons circle round the atomic nucleus, without falling into it? According to antenna physics in electrodynamics the rapid circular motion of the electrons must lead to the radiation of electromagnetic waves, and the kinetic energy of the electron would thereby be used up very quickly. Quite obviously, however, there are atoms which have a lifetime of more than a picosecond. Why?

© Springer-Verlag GmbH Germany 2017
D. Stauffer et al., *From Newton to Mandelbrot*, Graduate Texts in Physics,
DOI 10.1007/978-3-662-53685-8_3

Photoeffect. Why does one use red light in the photographic darkroom? According to classical electrodynamics the energy density $(E^2 + B^2)/8\pi$ varies continuously with the field strength, and with sufficiently strong light intensity one should be able to knock electrons out of metal, or initiate physico-chemical processes in the photographic emulsion, even with red light. In practice, however, light with frequency v appears to occur only in energy quanta of strength hv, with the very small Planck action quantum h. This energy packet hv for red light (low frequency v) is insufficient, whereas with violet light (twice as high a frequency) the energy threshold can be surpassed. Why is the energy quantified in this way?

Electron Microscope. In a light microscope it is well known that normally one can see no structure smaller than the wavelength of the light. Electrons, on the other hand, are point-like, so one should be able with them to see any structure, however small. The electron microscope, however, does not work this way: the electrons appear to behave there like waves with a wavelength which is smaller, the higher the electrical voltage in the electron microscope. Now, are the electrons particles, or are they waves?

Uncertainty. What does Heisenberg's Uncertainty Principle signify? Was Einstein right when he said: God does not gamble? Or is everything in life preordained only by probabilities? Why, of a great number of identical radioactive atoms, does this one decay soon, and another only after a longer time, as one can easily verify from the clicks of a Geiger counter?

According to our present understanding, reality is described by a complex wave function $\Psi = \Psi(r, t)$, which is linked with the probability of finding a particle at the position r at the time t. The physically measurable quantities, such as, e.g., the momentum of the particle, are values calculable from Ψ. An electron can, according to the experimental arrangements, be described by a plane wave $\Psi \sim \exp(iQr)$, a delta function $\Psi - \delta(r)$, or a function Ψ lying between these two extremes. In the first case the electron behaves like a wave, in the second like a particle, and in general its behaviour lies between the two extremes.

3.1.2   Mathematical Foundations

For those readers who do not remember linear algebra very well, a few fundamentals are here recalled, while the others can merely learn a new notation for well-known facts. In general the quantum mechanical part is the mathematically most exacting part of this presentation, while for the subsequent statistical physics in particular little mathematics is required.

In the d-dimensional space of complex numbers $z = a + ib$, $z* = a - ib$ is the complex conjugate of z, a is the real part Rez and b the imaginary part Imz of z. As in real numbers, a matrix f multiplies a vector Ψ, giving for the i-th component

$\sum_k f_{ik} \Psi_k$, now also written as $f|\Psi\rangle$. The scalar product of two vectors Φ and Ψ is now written as $\langle\Phi|\Psi\rangle = \sum_k \Phi_k^* \Psi_k$; we notice that $\langle\Psi|\Phi\rangle = \langle\Phi|\Psi\rangle^*$ because of the complex numbers. The norm $\langle\Psi|\Psi\rangle$ is always real and never negative; we usually set $\langle\Psi|\Psi\rangle = 1$. The Hermite conjugate matrix $f^\dagger$ of the matrix f has the matrix elements $f_{ik}^\dagger = f_{ki}^*$. If a matrix is Hermite conjugate to itself, $f_{ik} = f_{ki}^*$, it is called Hermitian. Just as most real matrices in physics are symmetric, most matrices in quantum mechanics are Hermitian. For Hermitian f we have $\langle\Phi|f\Psi\rangle = \langle f\Phi|\Psi\rangle$, and we denote both with the symmetric notation $\langle\Phi|f|\Psi\rangle$ for $\sum_{ik} \Phi_i^* f_{ik} \Psi_k$.

When $f|\Psi\rangle = \text{const.}|\Psi\rangle$, we call this constant an eigenvalue and the associated $|\Psi\rangle$ an eigenvector, the multiplication by the matrix f then signifies a change of length, but not of direction of this eigenvector. The eigenvalues of a Hermitian matrix are always real, and their eigenvectors can be chosen to be orthonormal: $\langle\Psi_i|\Psi_k\rangle = \delta_{ik}$ with the Kronecker delta: $\delta_{ik} = 1$ when $i = k$ and zero otherwise. Instead of Ψ_k we also write simply $|k\rangle$, so that $f|k\rangle = f_k|k\rangle$ is the eigenvalue equation. If $\Psi = |\Psi\rangle$ is to be arbitrary, we also write $|\rangle$. Examination answer papers consisting only of blank pages, however, are not always evaluated in this sense as correct solutions to any arbitrary problems. Experts understand by this notation, invented by Dirac, a general concept of quantum mechanics, which we simply ignore.

Matrices are multiplied according to the formula:

$$(f\,g)_{ik} = \sum_j f_{ij} g_{jk},$$

$fg|\Psi\rangle$ therefore means that g is first applied to the vector Ψ, then f to the result. We have $(fg)^\dagger = g^\dagger f^\dagger$; fg is usually different from gf. The commutator

$$[f, g] = fg - gf$$

of two matrices will be crucially important for quantum mechanics; if all commutators were null, there would be no quantum effects.

In *Hilbert space* we set the dimensionality $d = \infty$, and instead of summing the components over the index k in scalar products or multiplication by matrices, we integrate over a continuous k from $-\infty$ to $+\infty$. Accordingly we rename k as x, so that, for example,

$$\langle\Phi|\Psi\rangle = \int_{-\infty}^{\infty} \Phi^*(x)\Psi(x)\,dx.$$

where we have till now had matrices we now have linear operators, operating on the functions $|\Phi\rangle = \Phi(x)$; the best known operator is the gradient (the nabla operator ∇). The product of two operators corresponds again to the sequential execution of the two operations. The exponential function of an operator f is interpreted in the sense of a power series of the operator. Mathematicians can of course define Hilbert space much more precisely; I find it most convenient when in doubt to think of a finite matrix in d dimensions.

3.1.3   Basic Axioms of Quantum Theory

We shall not construct quantum mechanics here in a strictly axiomatic way: only somewhat more axiomatically than classical mechanics. We therefore start from a few postulates, with which we can go a long way; only very much later shall we need still further postulates. Our three basic axioms now are:

1. The state of an object is described by its wave function $\Psi = \Psi(x, t)$.
2. $|\Psi|^2$ is the probability that the object has time t and coordinate x.
3. The physically observable quantity f corresponds to the linear Hermitian operator f, so that

$$\bar{f} = \langle \Psi | f | \Psi \rangle$$

is the experimental average value ("expectation value") of this quantity; the individual measured values f_n for this quantity are eigenvalues of the operator f.

In these axioms x stands for the totality of all the coordinates, thus (x, y, z) for a particle. In this sense integration for a scalar product $\langle \Phi | \Psi \rangle$ is an integration over all the position variables x. We do not need any other than linear Hermitian operators for the characterisation of physical quantities; often the operator f is differentiated by a circumflex ˆ from the measured quantity f (e.g., f can be the momentum):

$$\bar{f} = \langle \Psi | \hat{f} | \Psi \rangle.$$

Because of the interpretation of $|\Psi|^2$ as a probability (more precisely: a probability density) Ψ must be normalised: $\langle \Psi | \Psi \rangle = 1$. When the wave function Ψ is an eigenfunction of the operator f, so that $f\Psi = f_n\Psi$, then only this one eigenvalue f_n occurs as a measured value, since $\langle \Psi | f | \Psi \rangle = \langle \Psi | f_n \Psi \rangle = f_n \langle \Psi | \Psi \rangle = f_n$, If, on the other hand, Ψ is not an eigenfunction, then in general all possible eigenvalues of the operator f occur in the measurement; the latter indicates the well-known indeterminacy of quantum mechanics. If the wave function is an eigenfunction of the operator, then there is no uncertainty in the measured value of f: f is sharply defined, all measured values are equal.

Let us consider, for example, a free particle in the volume V. As trial solution a plane wave:

$$\Psi = \text{const. } \exp(i(\boldsymbol{Q}\boldsymbol{r} - \omega t),$$

will describe the particle (this trial solution solves the Schrödinger equation to come later). This, after all, is where the name "wave function" comes from, although (sadly) Ψ is no longer a plane wave for a particle with interactions. The absolute value of this exponential function is 1, and so the normalisation

$$1 = \langle \Psi | \Psi \rangle = \int |\text{const.}|^2 1 \, d^3 r = |\text{const.}|^2 V,$$

yields const. $= 1/\sqrt{V}$. The particle is therefore to be found anywhere with equal probability.

Einstein and many others accepted this interpretation as probability either never or only reluctantly. The Schrödinger cat paradox appears to disprove the interpretation: if one locks up a cat Ψ in a cage V, then covers the cage with a cloth, and then introduces a dividing wall in the middle of the cage, the cat must be in one half or the other. Ψ, however, is equal in the two halves, since we do not know where the cat is. If one half is now sent to Australia, where does one put the cat food in? This question was once the subject of vehement discussion.

With what probability w_n does the eigenvalue f_n occur in a measurement of the quantity f? To answer this let us develop Ψ according to the orthonormalised system of the eigenfunctions Ψ_n of the operator f : $\Psi = \sum_n a_n \Psi_n$.

The expected value $\sum_n w_n f_n$ for f is then

$$\bar{f} = \langle \Psi | f | \Psi \rangle = \left\langle \sum_n a_n \Psi_n | f | \sum_m a_m \Psi_m \right\rangle = \sum_{nm} a_n^* a_m f_m \langle \Psi_n | \Psi_m \rangle$$
$$= \sum_n |a_n|^2 f_n,$$

so evidently $w_n = |a_n|^2$ is the required probability. The coefficients a_n can now be calculated by multiplying the defining equation $\Psi = \sum_n a_n \Psi_n$ scalarly by $\langle \Psi_m |$:

$$\langle \Psi_m | \Psi \rangle = \sum_n \langle \Psi_m | a_n | \Psi_n \rangle = \sum_n a_n \delta_{mn} = a_m.$$

Just as for the components of a three-dimensional vector we therefore have

$$\Psi = \sum_n \Psi_n \langle \Psi_n | \Psi \rangle \quad \text{or} \quad | \rangle = \sum_n |n\rangle \langle n|$$

for arbitrary $\Psi = | \rangle$, or in short: the sum $\sum |n\rangle \langle n|$ is the unit operator. Moreover, the Fourier transformation is nothing but just this trick with $\Psi_n \sim \exp(i n x)$, where n is then the wave vector. (In the Summary below we write x instead of $\boldsymbol{x}$ for simplicity, as if the world is one-dimensional).

Summary: Mathematical Formulae for Quantum Mechanics

Scalar product $\quad\quad \langle \Phi | \Psi \rangle = \int \Phi^*(x)\Psi(x)\mathrm{d}x$

Normalisation $\quad\quad 1 = \langle \Psi | \Psi \rangle = \int \Psi^*(x)\Psi(x)\mathrm{d}x$

Expectation value $\quad \bar{f} = \langle \Psi | \hat{f} | \Psi \rangle = \int \Psi^*(x)\hat{f}\Psi(x)\mathrm{d}x$

Expansion $\quad\quad\quad \Psi = \sum_n \Psi_n \langle \Psi_n | \Psi \rangle$ or $|\rangle = \sum_n |n\rangle\langle n|\rangle$

Orthonormalisation $\langle n|m \rangle = \int \Psi_n^*(x)\Psi_m(x)\mathrm{d}x = \delta_{nm}$ $\quad\quad$ (3.1)

Eigenvalue $\quad\quad\quad \hat{f}\Psi_n = f_n \Psi_n$

Hermitian $\quad\quad\quad \hat{f}^\dagger = \hat{f}$ or $\langle \Psi | \hat{f}\Phi \rangle = \langle \Psi | \hat{f} | \Phi \rangle = \langle \hat{f}\Psi | \Phi \rangle$

Probability $\quad\quad\quad w_n = |\langle \Psi_n | \Psi \rangle|^2 = \left| \int \Psi_n^*(x)\Psi(x)\mathrm{d}x \right|^2$.

When can two different measured quantities f and g both be sharply defined, and not only for a particular Ψ, but for all Ψ? According to the above rules this requires that all eigenfunctions of the operator f be also eigenfunctions of the operator $g : f|n\rangle = f_n|n\rangle$ and $g|n\rangle = g_n|n\rangle$ with the same eigenfunctions $\Psi_n = |n\rangle$. We therefore have $\Psi = \sum |n\rangle\langle n|\Psi\rangle$ for the commutator:

$$[g,f]\Psi = \sum_n (gf - fg)|n\rangle\langle n|\Psi\rangle = \sum_n (g_n f_n - f_n g_n)|n\rangle\langle n|\Psi\rangle = 0$$

for every Ψ, and hence $[g, f] = 0$: the two operators f and g must be interchangeable; then and only then are the two associated measured quantities sharply defined at the same time, and one has a common system of eigenfunctions. Heisenberg's uncertainty principle will tell us how large the two uncertainties are, when $[f, g]$ is not zero, but $\pm i\hbar$.

As promised, we shall usually omit the symbol for operators, and also shall only seldom employ the $|n\rangle$ notation.

3.1.4 Operators

We need here only two operators; one of them is trivial:

$$\begin{aligned} \text{position operator} &\quad \hat{r}\Psi = r\Psi \\ \text{momentum operator} &\quad \hat{p}\Psi = -i\hbar\partial\Psi/\partial r = -i\hbar\nabla\Psi. \end{aligned} \quad (3.2)$$

The position operator is therefore multiplication by the position coordinates, the momentum operator is, to within a factor, the gradient. The other operators such as angular momentum $r \times p$ can be derived from them. Especially important is the Hamilton operator H, which represents the energy, expressed as a function of the position and momentum operators. For an individual free particle we have $H = p^2/2m$, or

$$\hat{H} = (-i\hbar\nabla)^2/2m = -(\hbar^2/2m)\nabla^2.$$

We need this form as the Laplace operator in quantum mechanics much more often than the original definition of the momentum operator.

The quantity $\hbar = h/2\pi$ is used more often than the old Planck action quantum h:

$$\hbar = 1.054 \times 10^{-27} \text{erg} \cdot \text{s}. \tag{3.3}$$

The eigenfunctions of the position operator should give a sharply defined position and are therefore delta functions:

$$\hat{r}\Psi = r_n\Psi, \quad \text{hence} \quad \Psi \sim \delta(r - r_n)$$

with arbitrary eigenvalue r_n. Eigenfunctions of the momentum operator p must satisfy

$$-i\hbar\nabla\Psi = p\Psi, \quad \text{hence} \quad \Psi \sim \exp(ipr/\hbar)$$

so the eigenfunctions of the momentum are plane waves. The wave vector Q has been coupled to the momentum eigenvalue p since 1923 by Louis de Broglie's (1892–1987) equation:

$$p = \hbar Q \tag{3.4}$$

wholly analogous to

$$E = \hbar\omega, \tag{3.5}$$

viz. Einstein's relation (1905) between energy E and frequency ω (see below); they can also be written as $p = h/\lambda$ and $E = hv$.

Particles with a sharp momentum are therefore described by plane waves in Ψ. If, on the other hand, they have a fixed position, then Ψ is a delta function. The particle-wave dualism therefore comes from the fact that the wave function Ψ of a particle is, according to the experimental set-up, sometimes more of a delta function (with sharp position) and sometimes more of a plane wave (with sharp momentum); usually it is neither of these two extreme cases. Position and momentum can, from this consideration, scarcely be both sharply defined at the same time, and this is clear also from the commutator for one dimension

$$[x, p]\Psi = -i\hbar\left[x, \frac{\partial}{\partial x}\right]\Psi = -i\hbar\left(x\frac{\partial\Psi}{\partial x} - \frac{\partial(x\Psi)}{\partial x}\right) = i\hbar\Psi,$$

or

$$[x_i, p_k] = i\hbar\delta_{ik} \tag{3.6}$$

in three dimensions.

3.1.5   Heisenberg's Uncertainty Principle

In the case when the operators f and g are conjugate to each other, i.e. when their commutator $[f, g] = \pm i\hbar$, then we have for the associated uncertainties Δf and Δg:

$$\Delta f \Delta g \geq \hbar/2. \tag{3.7}$$

Here $(\Delta f)^2 = \langle \Psi | (\hat{f} - \bar{f})^2 | \Psi \rangle$ is the expected value, in the sense of (3.1), of the mean square deviation, familiar from error estimation. This uncertainty relation (Heisenberg, 1927) is perhaps the fundamental deviation of quantum mechanics from classical mechanics.

Its proof uses the Schwarz inequality of linear algebra,

$$\langle \Psi | \Psi \rangle \langle \Phi | \Phi \rangle \geq |\langle \Psi | \Phi \rangle|^2,$$

which for three-dimensional vectors is trivial: $|\Psi \Phi| \geq |\Psi \Phi \cos \alpha|$ with angle α between Ψ and Φ. We use

$$F = \hat{f} - \bar{f}, \ G = \hat{g} - \bar{g}, \ \Psi_1 = F\Psi, \ \Psi_2 = G\Psi.$$

Hence

$$(\Delta f)^2 (\Delta g)^2 = \langle \Psi | FF | \Psi \rangle \langle G\Psi | G\Psi \rangle = \langle \Psi_1 | \Psi_1 \rangle \langle \Psi_2 | \Psi_2 \rangle \geq |\langle \Psi_1 | \Psi_2 \rangle|^2$$

and

$$\langle \Psi_1 | \Psi_2 \rangle = \langle \Psi | FG | \Psi \rangle = \langle \Psi | [F, G] + GF | \Psi \rangle$$
$$= \pm i\hbar + \langle \Psi | GF | \Psi \rangle = \pm i\hbar + \langle \Psi_1 \langle \Psi_2 \rangle^*$$

corresponding to the assumption that $[f, g] = \pm i\hbar$. Accordingly $\pm i\hbar/2$ is the imaginary part of $\langle \Psi_1 | \Psi_2 \rangle$. The modulus of a complex number is never smaller than the modulus of its imaginary part: $|\langle \Psi_1 | \Psi_2 \rangle| \geq \hbar/2$, or $\Delta f \Delta g \geq \hbar/2$, as asserted in (3.7). An electron localised to 1 Å has therefore a momentum uncertainty of at least $\Delta p = \hbar/2\Delta x$, which corresponds to a velocity of one thousandth of the velocity of light. Accordingly we can at first neglect relativistic effects in atomic structure. The heavier a particle is, the less is its uncertainty; he who parks his Rolls Royce in a No Parking spot cannot appeal to Werner Heisenberg (1901–1976).

Anyone who finds the above derivation too formal may instead take $\Psi(x)$ in one dimension as the Gauss curve $\exp(-x^2/2\sigma^2)$ and then Fourier transform. The Fourier components as function of the wave vector Q again form a Gauss curve with a width proportional to $1/\sigma$. The width σ in position space therefore corresponds to the width $1/\sigma$ in wave vector space or the width $\hbar/\sigma$ in momentum space ($p = \hbar Q$). For $\sigma \to 0$ we obtain a delta function at the position ("particle", constant in the momentum space); for $\sigma \to \infty$ we obtain a constant in position space ("wave", delta function in the momentum space). In general the Gauss function lies between the two extremes of particle and wave.

3.2 Schrödinger's Equation

3.2.1 The Basic Equation

Quantum mechanics is based on the time-dependent Schrödinger equation

$$\hat{H}\Psi = i\hbar\frac{\partial\Psi}{\partial t} \tag{3.8}$$

with the Hamilton operator H; this is the energy, described as a function of momentum (operator) and position. We postulate it here as a further basic axiom, but one can make it comprehensible if one believes with Einstein that "energy/h = frequency". Since the Hamilton operator has the dimension of energy, the time derivative of the wave function on the right-hand side in (3.8) must be multiplied by h or $\hbar$, since derivation with respect to time gives the added dimension of frequency.

The only things left unclarified by this dimensional argument are the dimensionless factors. For example, for a single particle with potential energy $U(r)$ in three dimensions we have:

$$H = \frac{p^2}{2m} + U = \frac{-\hbar^2\nabla^2}{2m + U} \to \hbar^2\frac{\nabla^2\Psi}{2m} + U(r)\Psi = i\hbar\frac{\partial\Psi}{\Psi t}.$$

In this generality it is a matter of a linear differential equation with a variable coefficient $U(r)$; it can be solved on the computer. It is simpler if no forces are present: $U = 0$ and $-\hbar^2\nabla^2\Psi/2m = i\partial\Psi/\partial t$. The solutions here are plane waves:

$$\Psi \sim \exp(iQr - i\omega t) \quad \text{with} \quad \hbar\omega = \frac{\hbar^2 Q^2}{2m}; \tag{3.9}$$

The apparently so trivial factor i in front of $\partial\Psi/\partial t$ makes (3.8) drastically different from a diffusion or heat conduction equation and produces waves like the wave equation. Unlike the wave equation, however, $\omega \sim Q^2$, not $\omega = cQ$. In this example $\hbar\omega(=p^2/2m)$ is the energy, and this is true quite generally.

In nearly all applications it is not (3.8) that is solved, but the eigenvalue equation (see Sect. 3.1.2) for the Hamilton operator

$$\hat{H}\Psi = E\Psi \tag{3.10}$$

with the eigenvalue E, the energy of the Hamilton operator. If one has found such an eigenfunction Ψ, then its time dependence is, according to (3.8), quite simple: $\Psi \sim \exp(-i\omega t)$ with $\hbar\omega = E$, as already stated in (3.5). The time dependence of Ψ is therefore given according to Einstein, the position dependence has to be laboriously found from (3.10). We call (3.10) the time independent Schrödinger equation.

An important special case concerns the above-mentioned single particle in the potential $U(\mathbf{r})$:

$$-\hbar^2 \frac{\nabla^2 \Psi}{2m} + U(\mathbf{r})\Psi = E\Psi; \tag{3.11}$$

This is the form of the Schrödinger equation which we shall most often use. Problems with two or more particles with interacting forces are difficult or impossible to solve exactly, so typical presentations of quantum mechanics deal mainly with (3.11) which is often soluble. Quantum chemistry deals with the calculation of more complicated molecules by the Schrödinger equation, where the forces between the constituent atoms are critically important. In spite of the availability of large computers, however, one does not solve (3.10) directly, but first makes appropriate approximations, which fall outside the scope of this presentation.

If there are N particles, which exert *no* forces on each other, then one can solve the Schrödinger equation (3.10) by a trial product solution (since the Hamilton operator is now the sum of the Hamilton operators of the individual particles): $\Psi(\mathbf{r}_1, \ldots, \mathbf{r}_N, t)$ is the product of the solutions of (3.11) applying to each particle. Such separation procedures are common in mathematics.

In what follows we treat (3.11) for one particle in soluble cases, looking particularly for new effects which do not occur in classical mechanics. We notice here that Ψ is a continuous function; the gradient of Ψ is also continuous, so long as the potential energy U is finite.

3.2.2   Penetration

Let a potential step in one dimension (Fig. 3.1) be given by $U(x < 0) = 0$; $U(x > 0) = U_0$. Of the two cases $E < U_0$ and $E > U_0$ we consider only the more interesting one: $E < U_0$. "Classically", i.e. without quantum effects, no particles can then penetrate the potential step, in quantum mechanics this does occur.

Fig. 3.1 Potential profile at a step. Classically, all particles are reflected at the step; only in quantum mechanics do they penetrate a short distance

Both on the left and on the right we have to solve the Schrödinger equation

$$-\hbar^2\Psi''/2m + U\Psi = E\Psi$$

and then to join the two solutions continuously at $x = 0$.

	left		right
	$-\hbar^2\Psi''/2m = E\Psi$	Ansatz	$-\hbar^2\Psi''/2m = (E - U_0)\Psi$
	$\Psi = Ae^{iQx} + Be^{-iQx}$	for	$\Psi = ae^{\kappa x} + be^{-\kappa x}$
	$\hbar^2 Q^2/2m = E$	solution	$\hbar^2\kappa^2/2m = U_0 - E$

Now Ψ and the derivative Ψ' must be continuous at $x = 0$:

$$A + B = a + b, \quad iQ(A - B) = \kappa(a - b).$$

These two equations are still insufficient to determine the four unknowns (A, B, a, b). We also know, however, that $\langle\Psi|\Psi\rangle = 1$. Therefore Ψ must not diverge exponentially as $x \to \infty$; so a must be zero. Moreover, we are interested more in the ratios B/A and b/A than the absolute values such as A (the latter can be determined if we specify the "volume" on the left of the step). The solution of $1 + B/A = b/A$ and $iQ(1 - B/A) = -\kappa b/A$ is

$$\frac{B}{A} = \frac{1 - i\kappa/Q}{1 + i\kappa/Q}, \quad \frac{b}{A} = \frac{2}{1 + i\kappa/Q}.$$

Classically one finds that $b = 0$ (no penetration), but quantum mechanically b differs from zero. The wave function can therefore penetrate as $e^{-\kappa x}$ into the region forbidden classically. The penetration depth, i.e. the region over which Ψ is still appreciably positive, is

$$\frac{1}{\kappa} = \frac{\hbar}{\sqrt{2m(U_0 - E)}}$$

and is the larger, the smaller the mass. If $\hbar$ were zero, the penetration depth would be zero. This is a special case of the general *Correspondence Principle*: the limiting case $\hbar \to 0$ gives back the classical mechanical solution. With an energy difference $U_0 - E$ of $1\,\text{eV}$ the penetration depth $1/\kappa$ for an electron is in the Angstrom region.

Fig. 3.2 Potential barrier with tunnel effect

It should be noticed that $|B/A| = 1$: all particles coming from the left (A) are reflected in the neighbourhood of the step and flow back towards the left (B). No particle stays for long in the forbidden region. In this sense the reflection coefficient, often denoted by R, is unity and the transmission coefficient $T = 1 - R$ is zero. For $E > U_0$ we have classically $R = 0$ and $T = 1$: all particles flow past the now inadequate potential threshold. Quantum mechanics, however, still gives a finite reflection probability R.

3.2.3 Tunnel Effect

Now we combine two potential steps of equal height to form a barrier, such as is shown in Fig. 3.2. Particles arrive from the left with an energy E below the energy of the potential barrier. Then, classically, no particles pass through the barrier, but the Schrödinger equation gives a finite wave function Ψ also on the right-hand side, as if the particles tunneled through the barrier. (In a similar way many students tunnel through their examinations, even if at the start of their studies the barrier seemed to them insuperable. This breakthrough, however, depends on work, not on $\hbar$).

The one-dimensional Schrödinger equation $-\hbar^2 \Psi''/2m = (E - U)\Psi$ is now to be solved in all three regions:

left	middle	right
$-\hbar^2\Psi''/2m = E\Psi$	$-\hbar^2\Psi''/2m = (E - U_0)\Psi$	$-\hbar^2\Psi''/2m = E\Psi$
$\Psi = Ae^{iQx} + Be^{iQx}$	$\Psi = \tilde{a}e^{\kappa x} + be^{-\kappa x}$	$\Psi = \alpha e^{iQx} + \beta e^{-iQx}$

where again we have: $\hbar^2 Q^2/2m = E$ and $\hbar^2 \kappa^2/2m = U_0 - E$. Since particles come only from the left, there is no negative wave vector on the right, i.e. $\beta = 0$. Again only the ratios of the amplitudes, such as B/A, are of interest; we determine these four unknowns from the continuity conditions (for Ψ and Ψ' at $x = 0$ and at $x = a$):

$$\Psi : A + B = a + b \quad \text{and} \quad ae^{\kappa a} + be^{-\kappa a} = \alpha e^{iQa}$$

$$\Psi' : iQ(A - B) = \kappa(a - b) \quad \text{and} \quad \kappa(ae^{\kappa a} - be^{-\kappa a}) = iQ\alpha e^{iQa}.$$

The solution is obtained after a little algebra (which could be left to a computer with algebraic formula manipulation):

$$\left|\frac{\alpha}{A}\right|^2 = \frac{4\lambda}{4\lambda + (e^{\kappa a} - e^{-\kappa a})^2(1+\lambda)^2/4} \quad \text{with} \quad \lambda = \frac{\kappa^2}{Q^2}.$$

For $\kappa a \to 0$ this tends to 1, for $\kappa a \to \infty$ to $16\lambda \exp(-2\kappa a)/(1+\lambda)^2$. Since the probabilities are always proportional to the square of Ψ, the transmission probability $T = 1 - R$ is just this square $|\alpha/A|^2$ of the amplitude ratio of outgoing to incoming waves. For large κa we therefore have

$$T \sim e^{-2\kappa a}. \tag{3.12}$$

One could, of course, have guessed this already. If the wave function Ψ with exp $(-\kappa x)$ penetrates a barrier, then this factor is $\exp(-\kappa a)$ at the end of the barrier of thickness a. Since, moreover, the transmission probability T is proportional to $|\Psi|^2$, (3.12) necessarily follows.

A practical application is the tunnel electron microscope, for which Binnig and Rohrer won a Nobel prize in 1986. Electrons tunnel from a surface into the open, when an electric field draws them out of a metal. Corrugations of the surface, such as are caused by individual atoms, modify the tunnel current exponentially according to (3.12), and hence are made visible.

3.2.4 Quasi-classical WKB Approximation

The simple result (3.12) has been generalised for an arbitrarily shaped potential barrier by the physicists Wentzel, Kramers and Brillouin (we of course ignore the contribution of the mathematicians). As in the approximation for integration in general, we can represent the barrier as a sum of many potential steps (Fig. 3.3), which we now assume to be infinitesimally small. At each step with thickness $a_i = dx$ the transmission probability is reduced by the factor $\exp(-2\kappa_i a_i)$, where again $\hbar^2 \kappa_i^2 / 2m = U_i - E$. The sum of all these steps gives the product

$$T = \prod_i T_i \sim \prod_i \exp(-2\kappa_i a_i) = \exp(-\sum_i 2a_i \kappa_i)$$

$$= \exp(-\int 2\kappa(x)dx);$$

or, setting

$$S = \int \hbar\kappa(x)dx = \int \sqrt{2m(U(x) - E)}dx$$

Fig. 3.3 Discrete
approximation to a potential
barrier for derivation of the
WKB approximation

Fig. 3.4 Potential well as
model for an electron near to
the atomic nucleus

we obtain

$$T \sim e^{-2S/\hbar} \quad (S \gg \hbar) \tag{3.13}$$

for the transmission probability. The integration goes only over the classically forbidden region $E < U(x)$. This approximation is called quasi-classical, because it is
valid only when $\hbar \ll S$.

3.2.5   Free and Bound States in the Potential Well

In Sect. 3.3 we shall learn that the quantum theory of the atom is mathematically
quite complicated. As a simple one-dimensional approximation for an atom, which
yet has many of the correct characteristics, we consider here an inverted potential
barrier, the potential well of Fig. 3.4: $U(x) = 0$ for $x < 0$ and $x > a$, $U(x) = -U_0$
for $0 < x < a$. If E is positive, the particle can come from infinity and pass on to
infinity. This corresponds to an electron, which is indeed scattered by the atomic
nucleus, but is not captured by it. More interesting are negative energies E, where
bound states occur, as we shall see.

As with the potential barrier (tunnel effect) we solve the Schrödinger equation:
on the left ($\Psi = A \exp(+\kappa x)$), in the middle ($\Psi = a \exp(iQx) + b \exp(-iQx)$) and
on the right ($\Psi = \beta \exp(-\kappa x)$); exponentially divergent components cannot arise.
Thus we have four unknowns; these occur in four equations from the continuity of Ψ
and $\Psi\prime$ at $x = 0$ and $x = a$. This system of homogeneous linear equations then has
a non-null solution only if the determinant of this 4×4 matrix is zero. Calculation[1]
shows that this determinant vanishes if

$$\tan \frac{Qa}{2} = \frac{-Q}{\kappa} \quad \text{or} \quad = \frac{+\kappa}{Q}.$$

---

[1] See, for example, E. Merzbacher, Quantum Mechanics (Wiley, New York, 1961), p. 102.

We solve these equations graphically by finding the intersections of the family of curves $y(Q) = \tan(Qa/2)$ with the curve $y(Q) = -Q/\kappa$ and with $y(Q) = +\kappa/Q$. (Here $\hbar\kappa = (-2mE)^{1/2} = (2mU_0 - \hbar^2 Q^2)^{1/2}$ is also a function of Q). There are in general a finite number of such intersection points, and this is the crucial point; details only confuse. (Anybody who has no desire to follow through all the calculations can satisfy himself with the well known fact that the determinant of a 4×4 matrix is a polynomial of the fourth degree in the matrix elements and has a maximum of four zeros. This can therefore scarcely give an infinite number of solutions).

This example has shown us for the first time that the Schrödinger equation can also have discrete solutions: only for certain values of Q and hence for certain values of the energy E is there a non-null wave function Ψ as a solution. The electron in this model can thus either be bound to the potential well with discrete energy values, or it can remain free with continuously variable energy. The electron behaves in just the same way in an actual atom. Either it can free itself from the atom ("ionisation"): the energy is then positive, but otherwise arbitrary. Or else it is captured by the atomic nucleus (bound state): then the energy can only take discrete negative values. In transition from one energy level to another, energy $\Delta E = \hbar\omega$ is liberated or used up, which leads to quite definite frequencies ω in the light spectrum of this atom.

Table salt, for example, always glows with the same yellow colour if one holds it in a flame. This colour is given by the energy difference ΔE between two discrete energy levels. By spectral analysis of this kind one can identify materials in distant stars or interstellar gas, without setting up a chemical laboratory there.

Instead of applying the above model to the electrons of an atom, one can also use it to study nuclear physics: the potential well then refers to the force binding together protons and neutrons in the atomic nucleus.

3.2.6 Harmonic Oscillators

The harmonic oscillator, which plays a prominent role in mechanics and electrodynamics, also does so in quantum mechanics. We therefore consider an individual particle of mass m in a one-dimensional potential $U(x) = Kx^2/2 = m\omega^2 x^2/2$, where ω is the classical frequency of the oscillator; damping is neglected. Accordingly we have to solve the Schrödinger equation

$$-\hbar^2 \Psi''/2m + m\omega^2 x^2 \Psi/2 = E\Psi.$$

We take this problem for the purpose of studying a computer simulation with the program OSCILLATOR.

In appropriate units the one-dimensional Schrödinger equation has the form $\Psi'' = (U - E)\Psi$, similar to Newton's equation of motion; acceleration = force/mass. We accordingly solve it in a similar way to that of the program given in Sect. 1.1.3b on mechanics. We use ps for Ψ, $p1$ for $d\Psi/dx$ and $p2$ for $\Psi'' = d^2\Psi/dx^2$. Let dx be the step length and $U = x$.

```
PROGRAM OSCILLATOR ▮▮▮▮▮▮▮▮▮▮▮▮▮▮▮▮▮▮▮▮▮▮▮▮▮▮▮▮▮
   10 e =1.1
   20 ps=1.0
   30 p1=0.0
   40 dx=0.01
   50 x=0.0
   60 x =x+dx
   70 p2=(x*x-e)*ps
   80 p1=p1+p2*dx
   90 ps=ps+p1*dx
   100 print ps,p1
   110 goto 60
   120 end
```

In the first line one inputs a trial value for the energy E, with the aim that $\Psi(x \to \infty)$ shall diverge neither towards $+\infty$ nor towards $-\infty$. (As soon as one is certain that a divergence is occurring one must abort the run, since the program is ignorant of this). After a few trial shots one finds that for $E = 1.005$ the wave function tends towards $+\infty$, and for $E = 1.006$ towards $-\infty$. Just as for the previous model of atomic structure there is accordingly a discrete energy eigenvalue near 1, for which a non-divergent solution exists; for values differing from this, Ψ diverges to infinity, so that the normalisation $\langle \Psi | \Psi \rangle = 1$ cannot be carried out. (That this value here appears to lie between 1.005 and 1.006 is a result of the finite step length dx. With shorter step length and better programs the computer calculates much more accurately). Before Ψ diverges it has the form of a Gauss curve. With high resolution graphics one can display this still better in a lecture room.

Long before there were computers, however, this Schrödinger equation had been solved analytically. With the dimensionless abbreviations

$$\varepsilon = \frac{2E}{\hbar\omega} \quad \text{and} \quad \xi = x\sqrt{\frac{m\omega}{\hbar}}$$

one obtains the form

$$-\frac{d^2\Psi}{d\xi^2} + \xi^2\Psi = \varepsilon\Psi,$$

which has a solution $\Psi_n \sim h_n(\xi) \exp(-\xi^2/2)$ which vanishes as $\xi \to \infty$ only when $\varepsilon_n = 2n + 1, n = 0, 1, 2$. Here the h_n are the Hermitian polynomials of n-th degree:

$$h_n(\xi) = \pm \exp(\xi^2) d^n \left[\exp(-\xi^2) \right] / d\xi^n.$$

Our computer program evidently gave the solution for $n = 0$, where h_n is a constant and Ψ_n is therefore proportional to $\exp(-\xi^2/2)$. (The solution $\varepsilon = 3$ for $n = 1$ is not found by the above program, since we then need $ps = 0.0$ and $p1 = 1.0$ as starting conditions when $x = 0$). One can read more about Hermite polynomials and the subsequently required Laguerre and Legendre polynomials in mathematical compilations.[2]

Of all these formulae the only really important one is $\varepsilon = 2n + 1$, or

$$E_n = \hbar\omega \left(n + \frac{1}{2} \right). \tag{3.14}$$

The energy of the harmonic oscillator is thus quantised in packets of $\hbar\omega = h\nu$, as recognised by Einstein in 1905 on the basis of Planck's formula for the energy of the radiation equilibrium. Even when $n = 0$ the energy is not zero, but $\hbar\omega/2$. This must be so because of the uncertainty relation: if E were zero, then the position x and also the momentum p would both be zero, and both would be sharply defined at the same time; but this is forbidden by Heisenberg. Accordingly $\hbar\omega/2$ is known as the zero-point energy.

All vibrations in physics have the property of the harmonic oscillator, that $E_n = \hbar\omega(n + 1/2)$ is quantised. This is true for sound waves just as it is for light waves. One calls n the number of quasiparticles for this vibration; if the energy is raised from E_n to E_{n+1}, and hence by $\hbar\omega$, then in this manner of speaking a quasiparticle is acquired. These particles, however, are not "real", since they have no mass and since the number of the quasiparticles is not constant: a vibration quantum of frequency 2ω can break down into two quasiparticles of frequency ω. These quasiparticles have names ending in "-on": phonons are the vibration quanta of sound waves, photons those of light waves, magnons those of magnetisation waves, plasmons those of plasma waves; also there are ripplons, excitons, polarons, gluons and other quasiparticles. Accordingly when an atom passes from a state of higher energy $\hbar\omega_1$ to one of lower energy $\hbar\omega_2$ it can lose the spare energy by emitting a light wave with frequency $\omega = \omega_1 - \omega_2$ a photon of energy $\hbar\omega$ is born.

If less than a million copies of this book are bought and its first author is therefore strung up, he will swing to and fro initially with frequency ω and energy $\hbar\omega(n + 1/2)$. Through friction, that is to say, through collisions with air molecules, this oscillator will slowly give off energy, without decreasing its frequency: n becomes smaller and phonons are annihilated. The mass of the author is, of course, so great that quantum effects can scarcely be measured: whether n is $1 + 10^{36}$ or only 10^{36} is hardly relevant; with these large numbers we can treat n approximately as a continuous variable. We see here again the correspondence principle at work: classical mechanics is the limiting case when $n \to \infty$ or $E \gg \hbar\omega$. Emotionally less interesting are the solids consisting of inert gases (like solid neon): the heavier the atom, the better can one calculate the lattice vibrations (phonons) by classical mechanics.

---

[2]M. Abramowitz, J.A. Stegun: Handbook of Mathematical Functions (National Bureau of Standards, Washington DC); I.S. Gradshteyn, I.M. Ryzhik: Tables of Integrals, Series, and Products (Academic Press, New York).

3.3   Angular Momentum and the Structure of the Atom

In order to pass from the one-dimensional examples treated so far to three dimensions, we must now consider the angular momentum operator, which does not exist in one dimension.

3.3.1   Angular Momentum Operator

Since in classical mechanics we have: angular momentum = position × momentum, and we have already met quantum mechanical position and momentum as operators, we do not now need any new definition of the angular momentum operator, but simply take $r \times (-i\hbar\nabla)$ as the angular momentum operator. It has the same dimensions as $\hbar$, so we shall use the dimensionless operator $\hat{L} = $ (angular momentum)$/\hbar$:

$$\hat{L} = -i r \times \nabla, \quad \hbar\hat{L} = r \times \hat{p}. \tag{3.15}$$

This angular momentum operator is connected with the Laplace operator ∇^2, which in spherical coordinates can be shown mathematically to be:

$$\nabla^2 \Psi(r, \vartheta, \varphi) = \frac{1}{r^2}\frac{\partial(r^2\partial\Psi/\partial r)}{\partial r} - \frac{1}{r^2}\hat{L}^2\Psi. \tag{3.16}$$

The quantum mechanical kinetic energy is therefore the sum of the rotational energy and the second derivative with respect to r:

$$-\hbar^2\frac{\nabla^2\Psi}{2m} = \left(\frac{\hbar^2}{2m}\right)\left(\frac{L^2}{r^2}\Psi - \frac{1}{r^2}\frac{\partial}{\partial r}\left(r^2\frac{\partial\Psi}{\partial r}\right)\right).$$

The z-component of the angular momentum, again transcribed from the mathematics of spherical coordinates, is particularly simple:

$$\hat{L}_z = -i\frac{\partial}{\partial\varphi}. \tag{3.17}$$

Therefore, just as the position dependence leads to the momentum, the angular dependence leads to the angular momentum.

It does not need mathematicians to calculate, with little effort, the commutators for the angular momentum. For example, $L_x L_y - L_y L_x = i L_z$, and generally

$$L \times L = iL \quad [L^2, L] = 0. \tag{3.18}$$

Here we have the cross product of a vector with itself, $L \times L$, naturally a commutator, since classically it is always zero. We have learnt earlier that two operators simultaneously describe measurable quantities sharply, if their commutator is zero, such as the square of the angular momentum and one of its three components, but not two components nor indeed three. Angular momentum therefore also obeys an uncertainty relation. Traditionally one takes the z-component as this one component, but this is only a question of notation. Accordingly the square of the modulus, and the z-component of the angular momentum can be sharply determined, whereas the x- and y-components are uncertain.

3.3.2 Eigenfunctions of L^2 and L_z

Since the square of the modulus and the z-component of the angular momentum operator commute with each other, they must have a common system of eigenfunctions,

$$\hat{L}^2 \Psi = \text{const.} \Psi \quad \text{and} \quad \hat{L}_z \Psi = l_z \Psi$$

with the two eigenvalues const and l_z. The z-component is easier to handle: $-i\partial \Psi / \partial \varphi = l_z \Psi$ is solved by $\Psi = \exp(il_z\varphi)$ according to (3.17). The wave function must be uniquely defined, so $\Psi(\varphi) = \Psi(\varphi + 2\pi)$; therefore the eigenvalue l_z is an integer: $l_z = m$ with $m = 0, \pm1, \pm2, \dots$. Again we have derived from a mathematical boundary condition a quantum effect, that the z-component of the angular momentum can change only in jumps of $\pm\hbar$.

The dependence of the eigenfunction on the other angle ϑ by (3.16) is more complicated, but is also known from mathematics:

$$\Psi = Y_{lm}(\vartheta, \varphi) \sim e^{im\varphi} P_{lm}(\cos\vartheta)$$
$$\hat{L}^2\Psi = l(l+1)\Psi; \quad \hat{L}_z\Psi = m\Psi; \quad l = 0, 1, 2, \dots, |m| \le l. \quad (3.19)$$

It is clear that the quantum number m of the z-component of the angular momentum cannot be larger than the quantum number l of the total angular momentum. It is to be noted that the square of the angular momentum has for quantum number, not simply l^2, but $l(l+1)$. Even when $m = l$ there is still some angular momentum for the x- and y-components, since otherwise all three components would be sharply determined in contravention of the uncertainty. Only as $l \to \infty$ does the difference between $l(l+1)$ and l^2 become negligible: the correspondence principle at high quantum numbers.

The Y_{lm} so defined are called "*spherical harmonics*", and the P_{lm} are the associated Legendre polynomials

$$P_{lm}(y) \sim (1-y^2)^{m/2} \left(\frac{d}{dy}\right)^{l+m} (y^2-1)^l.$$

(The m in P_{lm} is also often written as an upper index). The proportionality factor is chosen so that on integration over the whole solid angle Ω (hence over the whole spherical surface) the Y_{lm} are normalised:

$$\int Y_{lm}(\vartheta, \varphi) Y_{l'm'}(\vartheta, \varphi) d\Omega = \delta_{ll'} \delta_{mm'}.$$

The Y_{lm} are therefore very convenient as quantum mechanical wave functions, when the angular momentum is to be sharply defined. In atomic physics they accordingly play a prominent role, as we shall see immediately: the electron "orbits" the atomic nucleus with constant angular momentum and therefore has a wave function proportional to Y_{lm}.

3.3.3 Hydrogen Atom

Together with the harmonic oscillator the hydrogen atom represented a great success for quantum mechanics: the calculated formulae agreed with the measured values to great accuracy. The more complicated the atom is, the more difficult the calculation becomes. Here we make a compromise, by considering a single electron in the neighbourhood of an atomic nucleus with charge number Z (hence charge Ze) and we ignore the other $Z - 1$ electrons. The aim of the calculation is to determine the spectral lines ("colours of the atom"). For an arbitrary isotropic central potential $U = U(|r|)$, hence not only for the Coulomb potential $U = -Ze^2/r$, the Schrödinger equation $-\hbar^2 \nabla^2 \Psi / 2m + U\Psi = E\Psi$ can be solved by a product:

$$\Psi(r, \vartheta, \varphi) = R(r) Y_{lm}(\vartheta, \varphi).$$

In this trial separation of the variables the radial wave function R must, because of (3.16), fulfill the condition:

$$(\hbar^2/2m)\left[-r^{-2}d(r^2 dR/dr)/dr + r^{-2}l(l+1)R\right] = ER - UR.$$

The substitution $\chi(r) = rR(r)$ then leads to

$$-\hbar^2 \chi''/2m + \left[U + \hbar^2 l(l+1)/2mr^2\right]\chi = E\chi, \tag{3.20}$$

i.e. to a one-dimensional Schrödinger equation. In fact, the expression in square brackets is just the effective potential of classical mechanics, which we have come across before in (1.15). The Y_{lm} therefore reduce the three-dimensional problem to a one-dimensional Schrödinger equation, with $\chi(0) = \chi(\infty) = 0$ as boundary conditions.

If this one-dimensional Schrödinger equation (3.20) has the eigenvalues E_n, $n = 1, 2, \ldots$, then we so far have used three quantum numbers: n, l, m are always integers.

Fig. 3.5 Splitting of the energy levels for an angular momentum quantum number $l = 2$ in a small magnetic field

The energy, however, depends on n, not on m. For each l there are $2l + 1$ different m-values, and for each n there are again several l-values. Traditionally one uses letters rather than numbers for l. An f-electron has, for example, the angular momentum quantum number $l = 3$:

$$l = 0\ 1\ 2\ 3\ 4$$

$$\text{letters} \quad s\ p\ d\ f\ g$$

As soon as the angular momentum is different from zero, there is the Zeeman effect: the energies of the $2l + 1$ different wave functions for the same l become somewhat distinguishable if a small magnetic field is applied. The "orbiting" electron creates a magnetic dipole moment $\mu_B l$, with the Bohr magneton $\mu_B = e\hbar/2mc = 10^{-20}$ erg/Gauss (m = electron mass). Since the energy of a dipole in a field is given by—dipole moment times field, a magnetic field B (in z-direction) terminates the degeneracy, i.e. the $2l + 1$ different wave functions now acquire $2l + 1$ different energies $E_m - m\mu_B B$ with $-l \le m \le +l$ (Fig. 3.5). (This quantum number m has nothing to do with the mass m).

If we now take in particular the Coulomb potential $U(r) = -Ze^2/r$ for an atom, let us first make the equation

$$-\frac{\hbar^2}{2mr^2}\frac{d(r^2 dR/dr)}{dr} + \frac{\hbar^2}{2mr^2}l(l+1)R - \frac{Ze^2}{r}R = ER$$

dimensionless, just as we did for the harmonic oscillator. With $\xi = r/r_0$, $r_0 = \hbar^2/Zme^2 = 0.53\,\text{Å}/Z$ and $\varepsilon = 2E\hbar^2/Z^2 me^4$ we obtain

$$-\xi^{-2}\frac{d(\xi^2 dR/d\xi)}{d\xi} + \xi^{-2}l(l+1)R - \frac{2R}{\xi} = \varepsilon R. \tag{3.21}$$

For $Z = 1$ we get $r_0 = 0.53\,\text{Å}$ for the Bohr atomic radius and $me^4/2\hbar^2 = 1$ Rydberg $= 13.5\,\text{eV}$.

As with the program OSCILLATOR in Sect. 3.2.6, a computer can now calculate the wave function R. In this program ATOM we simply replace $x \times x$ in line 70 by the potential $-1/x$ and in the initial conditions (lines 20 and 30) put $ps = 0.0$ and $p1 = 1.0$ (the program now finds the function $\chi(r) = rR(r)$, see (3.20)). So we get the solutions without angular momentum. With energy E in line 10 close to -0.245 one obtains a function which increases at first and then smoothly sinks to zero, before it finally diverges. Sadly, the process is now less precise in the determination of the energy eigenvalue than it was in the case of the harmonic oscillator. If one takes account of the various factors of 2 in the dimensionless energy one finds that the ε above is four times the energy value in the program, so that for the exact solution one suspects that $\varepsilon = -1$. The program becomes somewhat more practical with a loop which prints out only every tenth value (angular momentum = 0).

```
PROGRAM ATOM
   10 e=-0.25
   20 ps=0.0
   30 p1=1.0
   40 dx=0.01
   50 x=0.0
   60 for i=1 to 10
   65 x=x+dx
   70 p2=(-1/x-e)*ps
   80 p1=p1+p2*dx
   90 ps=ps+p1*dx
   95 next i
  100 print ps,p1
  110 goto 60
  120 end
```

The suspected $\varepsilon = -1$ is confirmed exactly by mathematics: for a solution not diverging at infinity we must have $\varepsilon = -1/n^2$, with the principal quantum number $n = 1, 2, 3, \ldots$. Then

$$R \sim e^{-\xi/n}\xi^l L_{n-l-1}^{2l+1}(2\xi/n)$$

with a natural number n and an angular momentum quantum number $l = 0, 1, 2, \ldots, n - 1$. It is physically plausible that the angular momentum cannot be arbitrarily high if the energy quantum number n is fixed, since the rotation energy contributes to the total energy, with $n > l \geq m$. That the energy $\varepsilon = -1/n^2$ depends only on n and not on the angular momentum (l, m) must therefore not be misconstrued as the rotational energy being zero.

The associated Laguerre polynomials $L_{n-l-1}^{(2l+1)}$ are defined through the Laguerre polynomials L_{n+1}:

$$L_{n-l-1}^{(2l+1)}(x) = (-d/dx)^{2l+1}L_{n+1}(x)$$

with

$$L_k(x) = \sum_{i=0}^{k}(-1)^k x^k/k!,$$

which is here much less interesting than the total wave function

$$\Psi \sim \exp\,(-r/nr_0)\,r^l Y_{lm}\,(\vartheta,\ \varphi)\ \times\ (\text{polynomial in } r/nr_0) \tag{3.22}$$

and the energy

$$E = -\frac{Z^2 m e^4/2\hbar^2}{n^2} = -\frac{Z^2}{n^2} \cdot 13.5\,\text{eV} \tag{3.23}$$

of the electron. To a definite principal quantum number n there correspond n different angular momentum quantum numbers $l = 0, 1, \ldots, n-1$, and to each l there correspond $2l + 1$ different direction quantum numbers m: $-l \le m \le l$. Altogether there are $\sum_l (2l + 1) = n^2$ different wave functions with the same energy $\varepsilon = -1/n^2$: the degree of degeneracy is n^2. To these electron states bound to the atomic nucleus with energy < 0 there still correspond "scatter states" (Rutherford formula, scattering probability $\sim \sin^{-4}(\vartheta/2)$, see Sect. 3.4) where Ψ does not vanish at infinity but becomes a plane wave.

The differences $\hbar\omega_{12}$ between two energy levels therefore vary as $(1/n_1^2 - 1/n_2^2)$. For the hydrogen atom we set $n_1 = 1$, so we obtain for $m = 2, 3, \ldots$ the Lyman series; with $n_1 = 2$, $n_2 = 3, 4, \ldots$ gives the Balmer series; $n_1 = 3$ gives the Paschen series, etc. One "sees" these series as spectral lines of the atom: if an electron falls back from the energy $E_2 \sim 1/n_2^2$ to the energy $E_1 \sim 1/n_1^2$, it sends out a photon of frequency ω_{12}, with $\hbar\omega_{12} = E_2 - E_1$. The mathematical regularity observed in the series (the Balmer series lies in the visible region) was one of the motivations for developing quantum mechanics.

3.3.4 Atomic Structure and the Periodic System

We use now, and justify later, the Pauli principle: two electrons cannot coincide in all quantum numbers. Up to now we have come across three quantum numbers n, l and m, and since to each n there correspond just n^2 different wave functions (different l and m), the Pauli principle so far implies that only n^2 electrons can sit in an energy shell E_n. Accordingly the element helium (two electrons in the innermost shell $n = 1$) should not exist, and the next innermost shell ($n = 2$) should have only four of the actual eight elements (lithium, beryllium, boron, carbon, nitrogen, oxygen, fluorine, neon). Without oxygen, carbon and nitrogen life would be very difficult. Solution: the elementary particles are spinning.

In addition to the orbital angular momentum L, which we have already taken into account, the particle can also rotate about its own axis; this intrinsic angular momentum S is called the "spin". It also is quantised; the z-component of the spin can take only the values $-S, -S + 1, \ldots, S$, analogous to the quantum numbers m of the orbital angular momentum. Unlike the orbital angular momentum, however, S can also take half-integer values, and electrons, protons and neutrons all have in fact $S = \frac{1}{2}$. The z-component of the spin of these elementary particles is therefore either $+\frac{1}{2}$ (the spin points up) or $-\frac{1}{2}$ (the spin points down). We therefore have:

$$\text{degree of degeneracy} = 2n^2. \qquad (3.24)$$

The mechanistic representation of spin as rotation about its own axis is not correct, since according to present understanding electrons have no spatial extent. We have already encountered a similar conceptual difficulty concerning the tunnel effect (Sect. 3.2.3): electrons have no shovels with which to dig holes. Nevertheless both visualisations, tunnels and spinning, are useful aids to the understanding. Here we dispense with the derivation of spin from Dirac's relativistic generalisation of the Schrödinger equation and treat spin as experimentally justified. Its magnetic moment is about twice as great as that of the orbital angular momentum:

$$\text{magnetic moment} = 2\mu_B S. \qquad (3.25)$$

Now the Pauli principle allows us two electrons in the innermost shell (hydrogen and helium), 8 in the second and 18 in the third shell. However, because of the forces between the electrons, which are here neglected, the number 8 also plays an important role in the third shell and with still greater n-values: after neon there follow first the 8 elements Na, Mg, Al, Si, P, S, Cl and Ar, as each electron is added to the previous ones. The fourth shell, however, is then started with K and Ca before the ten missing electrons of the third shell are filled in, e.g., with iron.

The period 8, therefore, extends throughout the whole periodic table of the elements. As the outermost shell of electrons is of predominant importance for chemical effects, all elements in the same position of the 8-period have similar chemical properties. This is true, for example, for all elements with just one electron in the outermost shell (H, Li, Na, K, Rb, Cs, Fr); these combine very readily, as in table salt NaCl, with an element which has 7 electrons in its outermost shell: by displacement of one electron, both atoms achieve a closed and energetically stable outermost shell. He, Ne, Ar, Kr, Xe and Rn already have closed outer shells and therefore are very reluctant to enter into a compound; because of this exclusive behaviour they are called inert gases. All these chemical facts follow from our quantum mechanics and the Pauli principle. Now we shall explain how the latter arises.

3.3.5 Indistinguishability

When an electron orbits an atomic nucleus 1 with a wave function $\Psi_a(r_1)$, and another electron another atomic nucleus 2 with $\Psi_b(r_2)$, then one can describe the total system $\Psi(r_1, r_2)$ by a product trial solution, $\Psi = \Psi_a(r_1)\Psi_b(r_2)$, if there are no forces acting between the two atoms. This product trial solution is then a solution of the Schrödinger equation $(H_1 + H_2)\Psi = (E_1 + E_2)\Psi$, since the Hamilton operator H_1 acts only on the electron at r_1, and H_2 only on that at r_2. This solution, however, is not the unique solution, since all electrons are the same and are not distinguished by name. Because of this indistinguishability of the electrons, $\Psi_a(r_2)\Psi_b(r_1)$ is an equally good solution: electron 1 has exchanged places with electron 2 and, since it carries no driver's license, quantum mechanics is unaware of the exchange. If there are two particular solutions of a linear equation, then the general solution is a linear combination of the two:

$$\Psi(r_1, r_2) = A\Psi_a(r_1)\Psi_b(r_2) + B\Psi_a(r_2)\Psi_b(r_1).$$

Because of the indistinguishability of the two electrons the probability cannot be altered by the exchange:

$$|\Psi(r_1, r_2)|^2 = |\Psi(r_2, r_1)|^2, \text{ and hence } \Psi(r_2, r_1) = u\Psi(r_1, r_2)$$

with a complex number u of modulus 1. Because of the same indistinguishability, repeated exchanges must again alter the wave function by the factor u: $\Psi(r_1, r_2) = u\Psi(r_2, r_1)$. Hence $u^2 = 1$, and there are only two possibilities: in the exchange the wave function does not alter at all ($u = 1$, *symmetric*), or it alters only by a change of sign ($u = -1$, *antisymmetric*):

$$\begin{aligned} u = 1 &: \quad \Psi \sim \Psi_a(r_1)\Psi_b(r_2) + \Psi_a(r_2)\Psi_b(r_1) \\ u = -1 &: \quad \Psi \sim \Psi_a(r_1)\Psi_b(r_2) - \Psi_a(r_2)\Psi_b(r_1). \end{aligned} \tag{3.26}$$

Experimentally it is observed that particles with integer spin (e.g., spin = 0), known as *Bose* particles, behave in the one fashion, and particles with halfinteger spin (e.g., spin = $\frac{1}{2}$), known as *Fermi* particles, in the opposite fashion. We therefore have:

For 2 Fermions	with parallel spins	Ψ is antisymmetric.
For 2 Fermions	with antiparallel spins	Ψ is symmetric.
For 2 Bosons	with parallel spins	Ψ is symmetric.
For 2 Bosons	with antiparallel spins	Ψ is antisymmetric.

$$(3.27)$$

Anybody for whom that is too complicated may regard the total wave function as the product of a spin function and a position wave function. Then this total wave function is symmetric for Bosons and antisymmetric for Fermions. I find (3.27) more practical (Enrico Fermi from Italy, 1901–1954; Satendra Nath Bose from Bengal, 1894–1974).

The Pauli principle is now quite trivial: if two Fermions have the same position wave function, $\Psi_a = \Psi_b$ and their spins are parallel, then $\Psi(r_1, r_2)$ must be antisymmetric, and hence proportional to $\Psi_a(r_1)\Psi_a(r_2) - \Psi_a(r_2)\Psi_a(r_1)$ and therefore zero:

Two Fermions cannot coincide in all quantum numbers
and have parallel spins.   (3.28)

This principle of Wolfgang Pauli (1900–1958) does not hold for Bosons. Of course, two electrons belonging to *different* atoms can have the same n, l, m and parallel spins, because then the wave functions Ψ_a and Ψ_b are different.

Now who obeys Pauli and who does not? Electrons, protons and neutrons all have spin $= \frac{1}{2}$, are Fermions and obey the Pauli principle. Otherwise one could scarcely explain the conductivity of metals at room temperature. Pions, on the other hand, are Bosons, as are photons and phonons. Atomic nuclei composed of an even number of Fermions have integer spin, for an odd number the spin is half-integer. Thus ^{3}He belongs to the Fermions, and the much more common ^{4}He to the Bosons. In statistical physics (Chap. 4) we shall come to recognise the drastic difference in the behaviour of the two helium isotopes at low temperatures: ^{3}He obeys (roughly) the Pauli principle, whereas ^{4}He disregards it and instead undergoes Bose-Einstein condensation.

3.3.6 *Exchange Reactions and Homopolar Binding*

A piece of table salt sticks together, then, because the sodium gives one electron to the chlorine atom, and the two atoms are thus bound together by electrostatic forces. This sort of chemical bond is called heteropolar. The air molecules N_2 and O_2, like the organic bonds between carbon atoms, cannot be explained in this way. Here instead electrons belonging to two atoms act as a bonding cement holding the molecule together (not unlike many families). We shall study this so-called homopolar type of bond using the example of the hydrogen molecule H_2.

The Hamilton operator of the H_2 molecule (2 protons and 2 electrons) is

$$\hat{H} = \frac{p_1^2}{2m} + U(r_1) + \frac{p_2^2}{2m} + U(r_2) + V(|r_1 - r_2|),$$

where we treat the two protons as fixed in space because of their very much greater mass. (Here we still have to add to the operator H the repulsion of the two atomic nuclei). The potential V arises from the repulsion of the two electrons, $V(r) = e^2/r$, and will now be assumed so small that the wave functions of the two electrons (3.22) are not significantly altered by V. The remaining four terms of the above Hamilton operator are denoted by H_0, hence as the unperturbed operator. Then the energy is

$$E = E_0 + \Delta E = \langle \Psi | H | \Psi \rangle = \langle \Psi | H_0 | \Psi \rangle + \langle \Psi | V | \Psi \rangle$$

Fig. 3.6 Binding energy
(total energy minus sum of
the individual energies) for
the hydrogen molecule
(schematic). The upper curve
applies for parallel electron
spins, the lower for
anti-parallel. Only in the
latter case is there a stable
homopolar binding due to
the energy minimum at
distance $r = 0.8\,\text{Å}$
(experimentally: $0.75\,\text{Å}$)

with the unperturbed energy E_0 as the sum of the energies according to (3.22) and
the energy correction

$$\Delta E = \langle \Psi | V | \Psi \rangle. \tag{3.29}$$

If one substitutes the wave functions $\Psi_a(r_1)\Psi_b(r_2) \pm \Psi_a(r_2)\Psi_b(r_1)$ from (3.26) by the
wave functions for $n = 1$, and takes account of the trivial energy from the repulsion
of the atomic nuclei, then after much calculation one gets the curves of Fig. 3.6: for
two parallel spins the binding energy is always positive and decreases monotonically
with increasing separation distance; for two anti-parallel spins the binding energy has
a minimum at about $0.8\,\text{Å}$. Qualitatively this difference is quite plausible: according
to the Pauli principle the two electrons may not have their other quantum numbers
equal, if they have the same ("parallel") spins, but they may if their spins are anti-
parallel. The closer the atomic nuclei move, the more effective this Pauli principle
becomes.

Quantitatively the energy correction ΔE can be split into a term A independent
of the spin orientation and a term $\pm J$, whose sign depends on the spins:

$$\Delta E = \langle \Psi_a(r_1)\Psi_b(r_2) \pm \Psi_a(r_2)\Psi_b(r_1) | V | \Psi_a(r_1)\Psi_b(r_2) \pm \Psi_a(r_2)\Psi_b(r_1) \rangle$$
$$= 2(A \pm J)$$

with "+" for anti-parallel and "−" for parallel spins. The "exchange interaction" J
arises mathematically from the product of a term with alternating sign and a term
with constant sign; these two products are of equal size, as one sees by exchanging
the two integration variables. We thus have:

$$A = \int \Psi_a^*(r_1)\Psi_b^*(r_2) V \Psi_a(r_1)\Psi_b(r_2) \mathrm{d}^3 r_1 \mathrm{d}^3 r_2$$

$$J = \int \Psi_a^*(r_1)\Psi_b^*(r_2) V \Psi_a(r_2)\Psi_b(r_1) \mathrm{d}^3 r_1 \mathrm{d}^3 r_2. \tag{3.30}$$

The normal term A can readily be understood as integration over the two probabilities $|\Psi_a(r_1)|^2$ and $|\Psi_b(r_2)|^2$, multiplied by the corresponding energy $V(r_1 - r_2)$. The exchange term J on the other hand arises from the exchange of the two coordinates r_1 and r_2 in the wave function. J is thus a combination of the quantum mechanical principle of indistinguishability with the classical Coulomb repulsion between the electrons.

Whether J appears in $\Delta E/2 = A \pm J$ with a plus sign or a minus sign is determined by the spin orientation. Whether J itself is positive or negative is not easily determined because of the complicated integral (3.30). If J is positive in a solid body, then the neighbouring electron spins are striving to be parallel to each other (minimum of energy); if $J < 0$, then anti-parallel neighbouring spins are preferred. In the case $J > 0$ we therefore have ferromagnetism, whereas $J < 0$ corresponds to antiferromagnetism. The elements iron, cobalt and nickel are ferromagnetic at room temperature. In antiferromagnetism, on the other hand, the lattice splits itself into two mutually penetrating sublattices, so that within each sublattice the spins are parallel, whereas spins from different sublattices are anti-parallel, like black and white on a chess board. This is most easily understood in one dimension: in ferromagnetism ($J > 0$) all the spins are upwards; antiferromagnetic spins ($J < 0$) are upwards at even positions, and downwards at odd positions (assuming that one has pointed the magnet in the right direction). An antiferromagnetic triangular lattice is therefore very "frustrated", i.e. the spin does not know which of its neighbours to take account of.

3.4   Perturbation Theory and Scattering

In the previous section on exchange interactions we have already come across a method of solving the quantum mechanics problem approximately, when one cannot achieve it exactly. We calculated an energy correction $\Delta E = \langle \Psi | V | \Psi \rangle$ under the assumption that the wave function Ψ is not appreciably different from the already known solution at $V = 0$, i.e. that V is sufficiently small. We shall now derive this method systematically, i.e. we treat the potential V as a small perturbation and develop a Taylor series in V. Then the perturbation potential V can either be constant (steady, or time-independent perturbation theory) or oscillating with a certain frequency (unsteady, or time-dependent perturbation theory).

3.4.1   Steady Perturbation Theory

We therefore start by assuming that the solution $H_0 \Psi_{0m} = E_{0m} \Psi_{0m}$ for the unperturbed Hamilton operator H_0 is already known with the various eigenfunctions Ψ_{0m} and the corresponding energy eigenvalues E_{0m}. For simplicity we assume that these eigenvalues are not degenerate, i.e. that different eigenvalues E_{0m} correspond to the different indices m. Now a further interaction V will be taken into account,

$H = H_0 + V$, which is so small that one can use a Taylor series in V and truncate it after one or two terms. We accordingly seek an approximate solution for $H\Psi_m = E_m\Psi_m$. (V may be an operator).

The new, still unknown, wave functions Ψ_n can be represented, like any vector in Hilbert space, as linear combinations of the old Ψ_{0m}, since the latter form an orthonormalised basis:

$$\Psi_n = \sum_m c_m \Psi_{0m},$$

so that

$$E_n \sum_m c_m \Psi_{0m} = E_n \Psi_n = (H_0 + V)\Psi_n = \sum_m c_m H_0 \Psi_{0m} + \sum_m c_m V \Psi_{0m}$$

$$= \sum_m c_m E_{0m} \Psi_{0m} + \sum_m c_m V \Psi_{0m}.$$

We now form the scalar product of this equation with $\langle \Psi_{0k}|$:

$$E_n \sum_m c_m \langle \Psi_{0k}|\Psi_{0m}\rangle = \sum_m c_m E_{0m} \langle \Psi_{0k}|\Psi_{0m}\rangle + \sum_m c_m \langle \Psi_{0k}|V|\Psi_{0m}\rangle$$

or

$$E_n c_k = c_k E_{0k} + \sum_m c_m V_{km}$$

with the matrix elements

$$V_{km} = \langle \Psi_{0k}|V|\Psi_{0m}\rangle = \int \Psi_{0k}^*(r) V(r) \Psi_{0m}(r) \mathrm{d}^3 r.$$

The analysis up to here has been exact: now we approximate. In the sum involving the matrix elements V_{km} we replace c_m by its value δ_{nm} for $V = 0$, as long as we are interested only in terms of the first order in V (notice that, for $V = 0$, $\Psi_n = \Psi_{0n}$):

$$(E_n - E_{0k})c_k = V_{kn}. \tag{3.31}$$

If we set $k = n$, since $c_n \approx c_{0n} = 1$, we obtain $E_n - E_{0n} = V_{nn}$ in agreement with (3.29). With $k \neq n$ we obtain $c_k = V_{kn}/(E_n - E_{0k}) \approx V_{kn}/(E_{0n} - E_{0k})$; with $\Psi_n = \Psi_{0n} + \sum' c_k \Psi_{0k}$ we thus obtain in this *first order perturbation theory*:

$$E_n = E_{0n} + V_{nn}$$
$$\Psi_n = \Psi_{0n} + {\sum}' V_{kn}\Psi_{0k}/(E_{0n} - E_{0k}) \quad \text{with} \tag{3.32}$$
$$V_{kn} = \langle \Psi_{0k}|V|\Psi_{0n}\rangle.$$

The summation sign with the prime indicates that one omits one term (here: $k = n$). Strictly speaking we have used this formula $\Delta E = V_{nn}$ not only for homopolar binding, but also previously for the Zeeman effect, since at that time we had certainly not explained how the magnetic field altered the wave functions, but simply assumed $\Delta E = -$(magnetic moment) $\times B$ with the magnetic moment taken from the theory without magnetic field B.

3.4.2   Unsteady Perturbation Theory

It is not the theory now that becomes time-dependent (in the past century it has changed little), but the perturbation V. The most important example of this is the photoeffect, when an oscillating electric field E (potential $V = -eEx$) acts on an electron. What is the change now in the wave function Ψ_n, which we again expand as $\Psi_n = \sum_m c_m(t)\Psi_{0m}$?

Just as in the derivation of (3.31) we obtain from the time-dependent Schrödinger equation $i\hbar\partial\Psi_n/\partial t = (H_0 + V)\Psi_n$ now used here for the first time

$$i\hbar\frac{dc_k(t)}{dt} = \sum_m c_m(t)V_{km}(t)$$

as the exact result, from which, as $c_m \approx \delta_{nm}$ in the first approximation, it follows that

$$i\hbar\frac{dc_k(t)}{dt} = V_{kn}(t) \quad \text{or}$$

$$c_k = (-i/\hbar)\int V_{kn}dt. \tag{3.33}$$

If the perturbation extends only over a finite period of time, the integration from $t = -\infty$ to $t = +\infty$ converges, and the probability of passing from the state $|n\rangle$ to the state $|k\rangle$ is

$$|c_k|^2 = \hbar^{-2}\left|\int V_{kn}(t)dt\right|^2. \tag{3.34}$$

In general,

$$V_{kn} = \exp\left[i(E_{0k} - E_{0n})t/\hbar\right]\langle\Psi_{0k}(t = 0)|V(t)|\Psi_{0n}(t = 0)\rangle;$$

If the perturbation potential V is independent of time, then V_{kn} oscillates with frequency $(E_{0k} - E_{0n})/\hbar$. The integral over such an oscillation is zero if E_{0k} is different from E_{0n}. A time-independent perturbation can therefore never alter the energy of the perturbed object.

It becomes more interesting when the perturbation potential oscillates with $\exp(-i\omega t)$ as in the photoeffect: $V(r, t) = v(r)\exp(-i\omega t)$. Then the system is originally unperturbed if this formula holds only for positive time, while $V(r, t) = 0$ for $t < 0$. Now it follows from (3.33) that

$$c_k = (-i/\hbar) \int v_{kn} \exp\left[i(\omega_{0k} - \omega_{0n})t - i\omega t\right] dt$$

with the time-independent matrix elements

$$v_{kn} = \langle \Psi_{0k}(r, t = 0)|v(r)|\Psi_{0n}(r, t = 0)\rangle, \quad \omega_{0k} = E_{0k}/\hbar.$$

This integration can readily be carried out, and the resulting transition probability is

$$|c_k|^2 = |v_{kn}|^2 \sin^2(\alpha t)/\hbar^2\alpha^2$$

with $2\alpha = \omega_{0k} - \omega_{0n} - \omega$ and $\sin(\phi) = (e^{i\varphi} - e^{-i\varphi})/2i$. For large t the function $\sin^2(\alpha t)/\alpha^2$ is approximated by $\pi t\delta(\alpha)$, so that with the calculation rule $\delta(ax) = \delta(x)/|a|$ we get

$$|c_k|^2 = (2\pi t/\hbar)|v_{kn}|^2\delta(E_{0k} - E_{0n} - \hbar\omega)$$

so the transition probability increases linearly with time (as long as it is not too large, and hence is still within the region of validity of our linear approximation). The transition rate $R(n \rightarrow k)$ from state $|n\rangle$ to the state $|k\rangle$ is thus $|c_k|^2/t$:

$$R(n \rightarrow k) = (2\pi/\hbar)|v_{kn}|^2\delta(E_{0k} - E_{0n} - \hbar\omega). \tag{3.35}$$

a formula known as the *Golden Rule* of quantum mechanics.

We recognise in the delta function the quantum mechanical version of energy conservation: the difference between the initial and final energies must agree exactly with the energy $\hbar\omega$ of the perturbation. In the photoeffect $\hbar\omega$ is thus the photo-energy, and the Golden Rule describes the probability of a photon causing the electron to pass from its initial energy E_{0n} to the higher energy E_{0k}. We therefore see that a proper theory of this photoeffect is quite complicated, even if the result appears very simple.

However, not only must the energy be conserved, but the matrix element v_{kn} must also not be zero, if a transition is to be possible. In atomic physics "selection rules" determine when the matrix element is exactly zero. One can explain much more simply why one cannot shift an electron by light radiation from a piece of metal in New York to one in Hawaii: v_{kn} contains the product $\Psi_{0k}^* \Psi_{0n}$ of the wave functions of the initial state and of the end state and is therefore zero if the two wave functions do not overlap. The Golden Rule thus prevents the telephone from being superfluous.

Fig. 3.7 Vectors n and k of the incoming and outgoing waves, scatter angle ϑ and momentum transfer Q in elastic scattering

3.4.3   Scattering and Born's First Approximation

In Grenoble (France) one can not only climb mountains and ski, but also scatter neutrons. For this purpose one uses a very high flux nuclear reactor, which does not produce energy, but serves research on the solid state and fluids. Numerous neutrons are created by the reactor, strike the sample under test and are scattered by it. How can one work back from the scattering pattern to the structure of the sample?

For this we apply (3.35) with $\omega = 0$ (elastic scattering) or with $\omega > 0$ (inelastic scattering). Far from the sample the neutrons etc. are described by plane waves, thus $\Psi_{0k}(r, t = 0) \sim \exp(ikr)$. With $Q = k - n$ as the difference between the wave vector of the outgoing wave $|k\rangle$ and the incoming wave $|n\rangle$, see Fig. 3.7, for a steady perturbation field v the matrix element

$$v_{kn} \sim \int e^{-iQr} v(r) \mathrm{d}^3 r \qquad (3.36)$$

is therefore proportional to the Fourier transform of the perturbation potential v. The scattering probability ("scattering cross section") therefore varies with the square of the Fourier component of $v(r)$ (Max Born 1926). The magnitude of the *momentum transfer* Q can be conveniently worked out from the scattering angle ϑ: $Q = 2k \sin(\vartheta/2)$, since the wave vectors k and n have the same length (v steady, energy conservation, elastic scattering, hence $|k| = |n|$). If the scattering potential is independent of angle, $v = v(|r|)$, the integration over spherical coordinates gives

$$v_{kn} \sim \int r \sin(Qr) v(r) \mathrm{d}r / Q,$$

which for a Coulomb potential $v(r) - 1/r$ leads to $v_{kn} \sim 1/Q^2$. The scattering probability thus varies as $\sin^{-4}(\vartheta/2)$, and so Born's first approximation in quantum mechanics gives just the same law as Rutherford's scattering formula in classical electrodynamics (which is why we did not derive the latter at the time). Rutherford applied it to describe the scattering of alpha particles by atomic nuclei; today's elementary particle physicists have at their disposal higher energies and quarks.

If now the scattering potential varies with time, then as in the time-dependent perturbation theory the scattering probability is proportional to the Fourier transform with respect to position and time:

$$\text{scattering cross-section} \sim |\int V(r, t) e^{-i(Qr - \omega t)} \mathrm{d}^3 r \mathrm{d}t|^2,$$

Fig. 3.8 Scattering intensity as a function of ω = energy transfer/$\hbar$, in inelastic neutron scattering in solid bodies

where $\hbar\omega$ is the difference between outgoing and incoming neutrons in what is now inelastic scattering. As in the theory of relativity we accordingly have a fully equal ranking of position and time: the one gives the momentum transfer $\hbar\mathbf{Q}$, the other the energy transfer $\hbar\omega$; the four-dimensional Fourier transform gives the scattering probability.

Instead of neutrons one can also scatter electrons or photons; the formulae remain the same. For example, Max von Laue discovered by scattering x-ray photons from crystals that these are periodic: the Fourier transform was a sum of discrete Bragg reflections. Electrons are mainly appropriate for the study of surfaces (LEED = low energy electron diffraction), since they are quickly absorbed in the interior. In inelastic scattering one finds a Lorentz curve $1/(1 + \omega^2\tau^2)$ for motion damped with $\exp(-t/\tau)$. Inelastic neutron scattering by damped phonons in solid bodies therefore gives, as a function of the frequency ω, a resonance peak at the phonon frequency, and the width of this peak is the reciprocal lifetime τ of the vibration, as sketched in Fig. 3.8.

Questions

Section 3.1

1. What effects show that classical mechanics is incomplete and must be extended by quantum mechanics?
2. Is $(f + g)^2 = f^2 + g^2 + 2fg$ for operators $\hat{f}$ and $\hat{g}$?
3. What is $\sum |n\rangle\langle n|$? Is $|n\rangle\langle n|$ different from $\langle n|n\rangle$?
4. How large is $\langle\Psi|\Psi\rangle$, and why is this so?

Section 3.2

5. Write down the Schrödinger equation for one particle in a potential U.
6. Why does it follow from 5 that $d\Psi/dx$ or $\nabla\Psi$ is continuous for a finite potential?

Section 3.3

7. Which energy levels are occupied in the ground state of fluorine, neon and sodium ($Z = 9, 10, 11$)? (Without electron-electron interactions).
8. What are the line series of Lyman, Balmer, Paschen, ...?
9. What is the difference between the atomic nuclei of helium 3 and helium 4?
10. How are exchange interactions and Coulomb interactions related?

Section 3.4

11. To what is the transition probability $n \to k$ proportional?
12. What measures elastic (and inelastic) neutron scattering?
13. What is a Lorentz curve?

Problems

Section 3.1

1. Show that Hermitian operators have only real eigenvalues; are their eigenvectors always orthogonal?
2. Show that the Fourier transform of a Gauss function $\exp(-x^2/2\sigma^2)$ is itself a Gauss function. What uncertainty relationship exists between the two widths $\Delta x = \sigma$ and ΔQ?
3. What are the normalised eigenfunctions Ψ_n and their eigenvalues f_n for the operator $f = c^2 d^2/dx^2$ in the definition interval $0 \le x \le \pi$, if the functions are to vanish at the edge? What is the time dependence of the eigenfunctions if $\partial^2 \Psi_n/\partial t^2 = f \Psi_n$? To which problem in physics does this exercise apply?

Section 3.2

4. Calculate the relationship between $d\bar{f}/dt$ and the commutator $[H, f]$, analogous to the Poisson brackets of mechanics.
5. What are the energy eigenvalues and eigenfunctions in an infinitely deep one-dimensional potential well ($U(x) = 0$ for $0 \le x \le L$ and $U(x) = \infty$ otherwise)?
6. A stream of particles flowing from left ($x < 0$) to right ($x > 0$) is partially reflected, partially transmitted by a potential step

$$U(x < 0) = 0, \quad U(x > 0) = U_0 < E.$$

Calculate the wave function Ψ (to within a normalising factor).

Section 3.3

7. Calculate (without normalising) the wave function Ψ of a particle in the interior of a two-dimensional circular potential well of infinite depth [$U(|r| < R) = 0$, and $U(|r| > R) = \infty$] with the trial solution $\Psi(r, \varphi) = f(r)e^{im\varphi}$.
   Hint: $\nabla^2 = \partial^2/\partial r^2 + r^{-1}\partial/\partial r + r^{-2}\partial^2/\partial \varphi^2$ in two dimensions. The differential equation $x^2 y'' + xy' + (x^2 - m^2)y = 0$ is solved by the Bessel function $y = J_m(x)$, with $J_0(x) = 0$ at $x \approx 2.4$.
8. Calculate the ground state energy of the helium atom (experimentally: $79\,\text{eV}$), treating the electron-electron interaction as a small perturbation.
   Hint: The polynomial in (3.22) is a constant for the ground state. Moreover:

$$\frac{\iint \exp(-|r_1| - |r_2|)|r_1 - r_2|^{-1}d^3 r_1 d^3 r_2}{\iint \exp(-|r_1| - |r_2|)d_3 r_1 d^3 r_2} = \frac{5}{16}.$$

Section 3.4

9. Calculate the photoeffect in a one-dimensional harmonic oscillator (electron with $U = m\omega_0^2 x^2/2$), and hence the transition rate from the ground state to the first excited state in the electric field $\sim e^{-i\omega t}$.

10. Calculate in the first Born approximation the neutron scattering cross-section in a three-dimensional potential $V(r < a) = U_0$ and $V(r > a) = 0$. Show that, for fixed $U_0 a^3$, the length a cannot be measured by neutrons whose wavelength is much greater than a.

Chapter 4
Statistical Physics

Abstract Taking Boltzmann's probability $\sim \exp(-E/kT)$ as the basic axion, we derive both classical thermodynamics of the 19th century as well as quantum statistics and critical phenomena from the 20th century.

This chapter deals with heat and its atomic foundation and description. We shall also consider more applied questions:

1. What is "temperature", and which of the numerous quantum states Ψ_n are actually realised?
2. How do we calculate, at least approximately, the properties of real materials, such as, for example, the specific heat C_V?
3. Where do macroscopic quantum effects occur, i.e. not only in the atomic region?

4.1 Probability and Entropy

4.1.1 Canonical Distribution

Almost the whole of statistical physics follows from the basic axiom:

> The quantum mechanical eigenstates Ψ_n of the Hamilton operator
> are realised in thermal equilibrium with the probability ρ_n:
> $$\rho_n \sim \exp(-\beta E_n) \qquad (4.1)$$
> with $\beta = 1/kT$; $\quad k = k_B = 1.38 \times 10^{-16}$ erg/K.

Here T is the absolute temperature measured in (degrees) Kelvin [K]; many authors set k_B, the Boltzmann constant, equal to unity and then measure the temperature in ergs, electron volts or other energy units.

© Springer-Verlag GmbH Germany 2017
D. Stauffer et al., *From Newton to Mandelbrot*, Graduate Texts in Physics,
DOI 10.1007/978-3-662-53685-8_4

Critically important in this basic axiom is the proportionality sign $\sim$: the probability ρ_n is not equal to the exponential function, but $\rho_n = \exp(-\beta E_n)/Z$, so that the sum over all probabilities is unity, as it should be:

$$Z = \sum_n \exp(-\beta E_n), \qquad (4.2)$$

where we sum the Hamilton operator over all eigenstates; Z is usually called the *partition function*. The thermal mean value is denoted by $\langle \cdots \rangle$ and is defined as

$$\langle f \rangle = \sum_n \rho_n f_n \qquad (4.3)$$

with the quantum mechanically measured values f_n, hence the eigenvalues of the associated operator. We often also need the thermal deviation

$$\Delta f = \sqrt{[\langle f^2 \rangle - \langle f \rangle^2]} = \sqrt{[(f - \langle f \rangle)^2]}. \qquad (4.4)$$

We do not need to know all the details of the quantum mechanics; it is usually sufficient to know that $\hbar Q$ is the momentum and $\hbar \omega$ the energy and that there are well-defined quantum states.

> Our quantum states are always the eigenstates Ψ_n of the Hamilton operator H. This is not necessary: one can also work with any other arbitrary basis of the Hilbert space, and then define ρ as the quantum mechanical operator $\exp(-\beta H)/Z$. The partition function Z is then the trace of the operator $\exp(-\beta H)$, so that $\mathrm{Tr}(\rho) = 1$. I avoid here the problems where one needs such a representation, and work with energy eigenstates in which the operator ρ is diagonal, with the diagonal elements ρ_n.

What plausibility can we find for the basic axiom (4.1)? First there is the barometric formula for height. Let us assume that the temperature T of the atmosphere is independent of height h (this assumption is appropriate for our argument here, but dangerous when mountain climbing in the Himalaya); moreover, let the pressure P, volume V and the number N of air molecules be connected by $PV = NkT$ (*classical ideal gas*). If h is increased by dh, the air pressure is decreased by the weight (per cm^2) of the air in a layer of thickness dh:

$$dP = -dh\, mg\frac{N}{V} \quad \text{or} \quad \frac{dN}{dh} = -\frac{mgN}{kT}.$$

So N/V and P decrease proportionally to $\exp(-mgh/kT)$. Since mgh is the potential energy, this result agrees with the exponential function of (4.1).

A second argument is more formal: water in a two-litre flask behaves as it does in two separate one-litre flasks, as regards its internal properties, since the contribution of the surface to its total energy is negligible. The product of the probability $\rho(L)$ for the left-hand litre and the probability $\rho(R)$ for the right-hand litre is therefore equal to the probability $\rho(L + R)$ for the two-litre flask. The energy E_{L+R} of the two-litre flask is equal to the sum $E_L + E_R$ of the one-litre energies. Accordingly, since the probability ρ depends only on the energy E, the equality $\rho(E_L + E_R) = \rho(E_L) \times \rho(E_R)$ must hold. This, however, is a characteristic of the exponential function. From this argument one also learns that (4.1) is valid only for large numbers of particles; for small numbers of molecules the surface effects neglected here become important.

In this argument we applied a fundamental result of probability calculations, that statistically independent probabilities multiply each other. Thus if half of the students have blue eyes and the other half have brown eyes, and if ten per cent of the students fail the examinations in Theoretical Physics at the end of the year, then the probability that a student has blue eyes and has failed is $0.5 \times 0.1 = 5\%$. For according to modern knowledge Theoretical Physics has nothing to do with eye colour. If half of the students work harder than average, and the other half less hard than average, then the probability that a student is working harder than average and nevertheless fails in the examinations is significantly smaller than 0.5×0.1. The two probabilities are now no longer independent, but correlated: try it out!

In addition to these two traditional arguments, now a modern one (and not normal examination material). We let the computer itself calculate the probability. We take the Ising model from Sect. 2.2.2 on electrodynamics (in matter). Atomic spins were there oriented either upwards or downwards, which was simulated in that program by $is(i) = 1$ or $is(i) = -1$, respectively. If all the spins are parallel to each other, then the energy is zero; each neighbouring pair of antiparallel spins makes a constant contribution to the energy. Therefore if, in the square lattice, a spin is surrounded by k anti-parallel neighbours, the resulting energy associated with this is proportional to k, and the probability is proportional to $\exp(-\text{const}.k)$. On pure geometry there are just as many possibilities for $k = 4$ as for $k = 0$ (we need only reverse the spin in the middle), and also there are just as many for $k = 3$ as for $k = 1$. Accordingly, if we determine the number N_k, telling how often k anti-parallel spins occur in the simulation, then in equilibrium the ratio N_4/N_0 must correspond to $\exp(-4 \times \text{const.})$ and the ratio N_3/N_1 to $\exp(-2 \times \text{const.})$ if the axiom (4.1) and the program are correct: $N_4/N_0 = (N_3/N_1)^2$. In fact one finds this confirmed, provided one lets the computer run long enough for the initial deviations from equilibrium to become unimportant. (Still more impressive is the calculation in a cubic lattice, where we then find that $N_6/N_0 = a^3$, $N_5/N_1 = a^2$ and $N_4/N_2 = a$.)

We therefore add to the program given in electrodynamics the following:

$$142\, kk = is(i) * [is(i-1) + is(i+1) + is(i-L) + is(i+L)] + 4$$
$$144\, n(kk) = n(kk) + 1$$

and finally print out $n(0), n(2), n(6), n(8)$. (For the sake of simplicity $kk = 2k$ here. If your computer does not like $n(0)$, add 5 instead of 4 in line 142.) Initially, for example, one chooses $p = 20\%$ of the spins upwards, the rest downwards, since for $p < 8\%$ there is spontaneous magnetisation, and the algorithm does not function nicely.

Accordingly we believe from now on that axiom (4.1) is true. A simple application is to the connection between fluctuations ΔE of the energy and the specific heat $\partial \langle E \rangle / \partial T$; the latter is the quantity of heat required to warm a unit mass of a material by one degree Celsius or Kelvin. We have the quotient rule of the derivatives:

$$kT^2 \partial \langle E \rangle / \partial T = -\partial \langle E \rangle / \partial \beta = -\partial \left[\Sigma_n E_n y_n / \Sigma_n y_n \right] / \partial \beta$$

$$= \left[\sum_n E_n^2 y_n \sum_n y_n - \sum_n E_n y_n \sum_n E_n y_n \right] / \left[\sum_n y_n \right]^2 \quad (4.5)$$

$$= \langle E^2 \rangle - \langle E \rangle^2 = (\Delta E)^2 = kT^2 C_V$$

with the abbreviation $y_n = \exp(-\beta E_n)$. The specific heat, often abbreviated as C_V, is thus proportional to the fluctuations in the energy. Similarly one can derive, with the canonical ensemble to be treated later, that the susceptibility is proportional to the magnetisation fluctuations, and the compressibility to the fluctuations in the number of particles. This rule is often used to determine the specific heat or the susceptibility in computer simulations.

4.1.2   Entropy, Axioms and Free Energy

Whoever uses a desk or a room for a long time notices that this always becomes untidy. In general Nature has the same tendency to go from an initial tidy state to an untidy final state. We therefore need a measure for the untidiness (disorder), and this measure is called "entropy". With nine scraps of paper originally arranged in a square, one can easily show experimentally that the disorder increases when one disturbs them by a draught. It is also used to measure information content (Shannon) or economic inequality (Theil).

Suppose we have two energy levels, of which the higher has g-fold degeneracy ($g \gg 1$), whereas the lower is not degenerate ($g = 1$). (That is to say, in the first state there are g different wave functions having the same energy, whereas in the second state there is only one wave function. We have come across such examples in the hydrogen atom.) Let the energy difference between the two levels be very much smaller than kT and therefore thermodynamically unimportant. If now all the quantum states with (approximately) the same energy are equally probable, then the upper level will be occupied g times more often than the lower. The number g or the entropy $S = k \ln(g)$ is thus a measure for the disorder of the end state: even if we start in the lower, non-degenerate level ($g = 1$, $S = 0$), our equilibrium state, after some time, is the higher level with the greater entropy.

Things become more complicated if the energy difference between the two levels is no longer very small. Then Nature will, on the one hand, make the energy as small as possible (lower level, non-degenerate), and on the other hand make the entropy as great as possible (upper level, highly degenerate). The compromise between these two contradictory aims is the "free energy" $F = \langle E \rangle - TS$: this F is a minimum in equilibrium at a given temperature, as we shall see in general later. The ratio between the two probabilities of occupying the upper and the lower energy levels, is therefore selected in Nature so that the corresponding mean value of the free energy becomes minimal.

Quantitatively this looks as follows: the entropy S is defined by

$$S = -k\langle \ln \rho \rangle = -k \sum_{\mathrm{n}} \rho_{\mathrm{n}} \ln \rho_{\mathrm{n}}, \tag{4.6}$$

where the distribution of the probabilities ρ_{n} in equilibrium is given by (4.1); one can, however, use (4.6) for an arbitrary distribution ρ_{n}, not corresponding to equilibrium. The sum is taken over all the different eigenfunctions Ψ_{n} of the Hamilton operator. If there is now a total of g different states Ψ_{n} in the energy interval between $\langle E \rangle - kT$ and $\langle E \rangle + kT$, then in equilibrium the main contribution to the sum in (4.6) comes from these g wave functions; since $1 = \sum_{\mathrm{n}} \rho_{\mathrm{n}} = g\rho_{\mathrm{n}}$ the probability for these states near $\langle E \rangle$ is therefore $\rho_{\mathrm{n}} = 1/g$, and so the entropy $S = -k \ln 1/g = k \ln g$, as stated above:

$$g = e^{S/k}. \tag{4.7}$$

In the derivation of (4.7) factors of order 1 are neglected, which add to the dimensionless entropy S/k terms of order 1. However, if each of the 10^{25} molecules in a glass of beer makes a contribution ≈ 1 to S/k, then this error does not matter: statistical physics is valid only for a large number of particles.

Why does one define S as a logarithm? One wants the entropy of two litres of water (see above example) to be equal to the sum of the entropies of the single litres of water:

$$
\begin{aligned}
S(L + R)/k &= -\langle \ln \rho(L + R) \rangle = -\langle \ln [\rho(L)\rho(R)] \rangle \\
&= -\langle \ln \rho(L) + \ln \rho(R) \rangle = -\langle \ln \rho(L) \rangle - \langle \ln \rho(R) \rangle \\
&= S(L)/k + S(R)/k,
\end{aligned}
$$

as desired. (Remark: it is always true that $\langle A + B \rangle = \langle A \rangle + \langle B \rangle$, whereas for $\langle AB \rangle = \langle A \rangle \langle B \rangle$ the two quantities A and B must be statistically independent.) The entropy is therefore an *extensive* quantity: it doubles with a doubling of the system, like, for example, energy and mass. The temperature and the pressure, on the other hand, are *intensive* quantities: they remain constant with doubling of the system. Quantities with behaviour intermediate between extensive and intensive are sometimes fractal, see Chap. 5.

It can now be shown, by a little algebra, that in thermal equilibrium at fixed energy $\langle E \rangle$ the entropy is maximal, and we have

$$\frac{dS}{d\langle E \rangle} = \frac{1}{T}. \tag{4.8}$$

The mathematical proof is left to the course teacher. From this follow the three axioms of thermodynamics:

1. In a closed system the energy is conserved.
2. In attaining equilibrium the entropy increases and is then maximal.
3. In equilibrium T is never negative. $\hfill (4.9)$

These hold for closed systems; if one supplies energy from outside, or tidies a desk, then $\langle E \rangle$ is not constant, and the entropy of the desk decreases. The third axiom, $T \geq 0$, is in reality more complicated ($S(T = 0) = 0$), which we shall here ignore. One approaches within 10^{-5} K of the absolute zero by adiabatic demagnetisation in order to measure properties of matter. If the temperature could be negative, then according to (4.1) arbitrarily high energies would occur with arbitrarily high probabilities, which is not the case. Half a century ago there was a fourth axiom under discussion, that there was a maximum temperature of 10^{12} K; this axiom, however, did not secure the necessary majority support.

So if one wanted to make coffee, one would continually raise the energy $\langle E \rangle$ of the water by dE, and hence the entropy by $dS = dE/T$, maintaining equilibrium all the time, thus heating it rather slowly. However, if rapid heating causes strong temperature gradients in the water, then it is not in equilibrium, and dS is greater than dE/T later (after turning off the heat), when the water passes into equilibrium: $dS \geq dE/T$. From this it follows that $dE \leq TdS$, since $T > 0$. The second axiom therefore also says that for a given entropy ($dS = 0$) the energy $\langle E \rangle$ becomes minimal when the closed system achieves equilibrium: $\langle E \rangle$ is minimal at fixed S, and S is maximal at fixed $\langle E \rangle$.

What we are really interested in, however, is equilibrium at fixed temperature, like the room temperature in an experiment. Then neither S nor $\langle E \rangle$ is constant, but only T. With the free energy $F = \langle E \rangle - TS$ already defined above we obtain the required extremal principle: $dF = dE - d(TS) = dE - T\,dS - S\,dT \leq T dS - T dS - SdT = -S\,dT$. Therefore if the temperature is constant ($dT = 0$), then $dF \leq 0$, when the system passes into equilibrium:

   − At fixed E, S is maximal in equilibrium.
   − At fixed S, E is minimal in equilibrium.
   − At fixed T, F is minimal in equilibrium. $\hfill (4.10)$

Now and later we usually omit the angled brackets for the energy $\langle E \rangle$ and other mean values, for in statistical physics we are nearly always dealing with the mean values.
  We have:

$$S/k = -\langle \ln \rho \rangle = -\langle \ln \exp(-\beta E_n)/Z \rangle$$
$$= \ln Z + \beta \langle E_n \rangle = \ln Z + E/kT, \quad \text{or}$$
$$-\ln Z = (E - TS)/kT = F/kT;$$

the partition function therefore is

$$Z = \sum_n \exp(-\beta E_n) = e^{-\beta F} = e^{S/k} e^{-\beta E} = g e^{-E/kT}. \tag{4.11}$$

Here again one sees the meaning of $g = e^{S/k}$ as "degree of degeneracy", hence as the number of different states with about the prescribed mean energy E.
  From quantum mechanics we need here first of all the information that there are discrete quantum states Ψ_n, over which we can, for example, sum to get the partition function Z; without this knowledge the degree of degeneracy g and the entropy are indeed not defined. In the solution of the Schrödinger equation we work with a fixed number of particles N in a fixed volume V. The derivative $dS/dE = 1/T$ in (4.8) is therefore really a partial derivative $\partial S/\partial E$ at constant V and N, which we make more precise by the notation $(\partial S/\partial E)_{VN}$. In the next section we generalise this concept so that, for example, we can also handle the equilibrium at constant pressure instead of constant volume, and can calculate not only the specific heat at constant volume but also at constant pressure.

4.2 Thermodynamics of the Equilibrium

The subject of this section is the classical thermodynamics of the 19th century, which was developed out of the theory of steam engines and can also be understood without quantum mechanics.

4.2.1 Energy and Other Thermodynamic Potentials

We have already come across $(\partial E/\partial S)_{VN} = T$, in reciprocal form. If one moves a piston of area A a distance dx into a cylinder with pressure P (as in an air pump), one performs the mechanical work $dE = PA dx = -P dV$, where dV is the volume change: $(\partial E/\partial V)_{SN} = -P$. Finally one defines $(\partial E/\partial N)_{SV} = \mu$; μ is called the chemical potential, a sort of energy per particle, to which we shall become accustomed.

If we therefore consider the energy E as a function of the extensive quantities S, V and N, the total differential is given by differentiating with respect to all three variables (in equilibrium):

$$dE = TdS - PdV + \mu dN. \tag{4.12}$$

We can, of course, also start from the entropy:

$$dS = (1/T)dE + (P/T)dV - (\mu/T)dN. \tag{4.13}$$

Other pairs of variables, whose product gives an energy, are velocity v and momentum $\boldsymbol{p}$, angular velocity $\boldsymbol{\omega}$ and angular momentum $\boldsymbol{L}$, electric field $\boldsymbol{E}$ and polarisation $\boldsymbol{P}$, magnetic field B and magnetisation M:

$$dE = TdS - PdV + \mu dN + v d\boldsymbol{p} + \boldsymbol{\omega} d\boldsymbol{L} + \boldsymbol{E} d\boldsymbol{P} + \boldsymbol{B} d\boldsymbol{M}. \tag{4.14}$$

When more than one sort of particles is present, we replace μdN by $\sum_i \mu_i dN_i$. Here $-PdV$ is the mechanical compression energy, $v d\boldsymbol{p}$ the increase in the translational energy, $\boldsymbol{\omega} d\boldsymbol{L}$ applies to the rotational energy, and the two last terms contribute to the electrical and magnetic energies, respectively. Then since $p^2/2m$ is the kinetic energy, a momentum change $d\boldsymbol{p}$ alters this energy by $d(p^2/2m) = (\boldsymbol{p}/m)d\boldsymbol{p} = v d\boldsymbol{p}$. One should not call μdN the chemical energy, since really no chemical reactions are involved; this term is the energy change from raising the number of particles. What then is TdS? This is the energy which does not belong to one of the named forms. For example, when we place a saucepan of water on the stove, then energy is fed into the water, without significantly altering (before it boils) V, N, $\boldsymbol{p}$, $\boldsymbol{L}$, $\boldsymbol{P}$ or $\boldsymbol{M}$. This form of energy is, of course, heat:

$$\text{quantity of heat } Q = TdS. \tag{4.15}$$

In this sense (4.14) represent the first axiom (energy conservation) in a particularly complete form. We always have an intensive variable (i.e. independent of the size of the system) combined with the differential of an extensive variable (i.e. proportional to the size of the system).

When one is considering ideal gases, the heat is the kinetic energy of the disordered motion. This is not true in general, however, for interacting systems: when ice is melting, or water turning into steam, we need a lot of additional heat to break the bonds between the H_2O molecules. This is rarely brought out in examinations, where after prolonged study, questions on heat theory produce only answers which are based on high school teaching on classical ideal gases and are not in general correct. Why, for example, is the specific heat at constant pressure greater than that at constant volume, and how does this come about in very cold water which, as is well known, contracts on heating?

If TdS is the measure of heat, then *isentropic* changes are changes at constant entropy without exchange of heat with the environment. One usually calls these

Table 4.1 Thermodynamic potentials with natural variables

Potential	Name	Differential	Natural variables
E	Energy	$dE = +TdS - PdV + \mu dN$	S, V, N
$E - TS = F$	Free energy	$dF = -SdT - PdV + \mu dN$	T, V, N
$E + PV = H$	Enthalpy	$dH = +TdS + VdP + \mu dN$	S, P, N
$E + PV - TS = G$	Free enthalpy	$dG = -SdT + VdP + \mu dN$	T, P, N
$E - \mu N$	–	$d(\ldots) = +TdS - PdV - Nd\mu$	S, V, μ
$E - TS - \mu N = J$	Gr. Can. Pot.	$dJ = -SdT - PdV - Nd\mu$	T, V, μ
$E + PV - \mu N$	–	$d(\ldots) = +TdS + VdP - Nd\mu$	S, P, μ
$E - TS + PV - \mu N$	–	$d(\ldots) = -SdT + VdP - Nd\mu$	T, P, μ

adiabatic (without transfer), since no heat passes through the walls. In *isothermal* changes, on the other hand, the temperature is constant.

By *Legendre* transformations similar to $F = E - TS$ one can now clarify quite generally which quantity is a minimum at equilibrium, for which fixed variables, as in (4.10). For example, $H = E + PV$ is a minimum at equilibrium, if entropy, pressure and number of particles are fixed. For $dH = dE + pdV + VdP = TdS + VdP + \mu dN$ is zero, if $dS = dP = dN = 0$. Starting from the energy, which is a minimum at equilibrium for fixed extensive variables, we can by this Legendre transformation form numerous other *thermodynamic potentials* (see Table 4.1) in addition to E, F and H. (Gr. Can. Pot. in the table abbreviates Grand Canonical Potential.) As long as we do not stir an electromagnetically heated cup of coffee in a railway train, we can concentrate on the three pairs of variables occurring in (4.12), and so arrive at $2^3 = 8$ different potentials. Each additional pair of variables doubles this number.

Natural variables are those which occur in the differential itself as differentials and not as pre-factors (derivatives); for fixed natural variables the corresponding potential is a minimum at equilibrium. F is also called the "Helmholtz free energy", and G the "Gibbs free energy".

One should not learn this table by heart, but be able to understand it: at the start we have the energy, with the extensive quantities as natural variables $dE = +\lambda d\Gamma + \cdots$. The Legendre transformed $E - \lambda\Gamma$ then has instead of the extensive variable Γ the corresponding intensive quantity λ as natural variable: $d(E - \lambda\Gamma) = -\Gamma d\lambda + \cdots$, and this trick can be repeated for each pair of variables. For the seven pairs of variables in (4.14) there are $2^7 = 128$ potentials, each with 7 natural variables. Each line of the table gives three derivatives, such as for example $(\partial J/\partial V)_T = -P$. Would you wish to learn them all by heart, particularly the 7×128 derivatives for the mentioned seven variable pairs?

If a saucepan with the lid on contains steam above and water below, the system is spatially inhomogeneous. If, on the other hand, a system is spatially homogeneous, and hence has the same properties overall, then the *Gibbs-Duhem* equation holds:

$$E - TS + PV - \mu N = 0. \qquad (4.16)$$

Proof All the molecules now have equal standing, so $G(T, P, N) = NG'(T, P)$ with a function G' independent of N. On the other hand $\mu = (\partial G/\partial N)_{TP}$; hence we must have $\mu = G' = G/N$, which leads us not only to (4.16), but also to a better interpretation of μ.

4.2.2   *Thermodynamic Relations*

How can one work out other material properties from properties already measured, without having to measure them also? For example, what is the difference between C_P and C_V, the specific heats at constant volume and constant pressure, respectively? We generally define the specific heat C as $T\partial S/\partial T$ and not as $\partial E/\partial T$, since it is concerned with the heat, not with the energy, which is needed for a rise of one degree in the temperature. Other material properties of interest are compressibility $\kappa = -(\partial V/\partial P)/V$, thermal expansion $\alpha = (\partial V/\partial T)/V$ and (magnetic) susceptibility $\chi = \partial M/\partial B$. In all the derivatives N remains constant, unless otherwise stated, and is therefore not explicitly written as a subscript: $\kappa_S = -(\partial V/\partial P)_{SN}/V$.

Purely by mathematical tricks of differentiation we can now prove exactly a great many thermodynamic relations:

(1) $\left(\dfrac{\partial x}{\partial y}\right)_z = \dfrac{1}{(\partial y/\partial x)_z}$

(2) $\left(\dfrac{\partial x}{\partial y}\right)_z = \left(\dfrac{\partial x}{\partial w}\right)_z \left(\dfrac{\partial w}{\partial y}\right)_z$ chain rule

(3) $\left(\dfrac{\partial}{\partial x}\right)_y \dfrac{\partial w}{\partial y} = \left(\dfrac{\partial}{\partial y}\right)_x \dfrac{\partial w}{\partial x}$ Maxwell relation: most important trick

(4) $\left(\dfrac{\partial x}{\partial y}\right)_z = -\left(\dfrac{\partial x}{\partial z}\right)_y \left(\dfrac{\partial z}{\partial y}\right)_x$ notice the sign

(5) $\left(\dfrac{\partial w}{\partial y}\right)_x = \left(\dfrac{\partial w}{\partial y}\right)_z + \left(\dfrac{\partial w}{\partial z}\right)_y \left(\dfrac{\partial z}{\partial y}\right)_x$

(6) $\dfrac{\partial(u, v)}{\partial(x, y)} = \dfrac{\partial(u, v)}{\partial(w, z)} \dfrac{\partial(w, z)}{\partial(x, y)}$, $\quad \dfrac{\partial(w, x)}{\partial(y, z)} \dfrac{\partial(s, t)}{\partial(u, v)} = \dfrac{\partial(w, x)}{\partial(u, v)} \dfrac{\partial(s, t)}{\partial(y, z)}$

$\dfrac{\partial(x, z)}{\partial(y, z)} = \left(\dfrac{\partial x}{\partial y}\right)_z = \dfrac{\partial(z, x)}{\partial(z, y)}$

Trick 4 follows from trick 5 with $w = x$, but is easier to learn. In trick 6 the *functional* or Jacobi determinants, which we also write as $\partial(u, v)/\partial(x, y)$, are defined as the 2×2 determinants: $(\partial u/\partial x)(\partial v/\partial y) - (\partial u/\partial y)(\partial v/\partial x)$; they also occur when the integration variables u and v in two-dimensional integrals are transformed into the integration variables x and y. All that, however, is not so important for us; what are crucial are the rules of trick 6, that one can calculate as if with ordinary fractions, and

that normal derivatives are special cases of these determinants. Now a few examples of tricks 1–6:
From (1) since

$$\left(\frac{\partial S}{\partial E}\right)_{VN} = \frac{1}{T}, \quad \left(\frac{\partial E}{\partial S}\right)_{VN} = T.$$

From (2)

$$\left(\frac{\partial E}{\partial T}\right)_V = \left(\frac{\partial E}{\partial S}\right)_V \left(\frac{\partial S}{\partial T}\right)_V = T\left(\frac{\partial S}{\partial T}\right)_V = C_V,$$

but

$$\left(\frac{\partial E}{\partial T}\right)_P = \left(\frac{\partial E}{\partial S}\right)_P \left(\frac{\partial S}{\partial T}\right)_P,$$

and this is not C_P.
From (3)

$$\left(\frac{\partial S}{\partial V}\right)_T = -\left(\frac{\partial}{\partial V}\right)_T \left(\frac{\partial F}{\partial T}\right)_V = -\left(\frac{\partial}{\partial T}\right)_V \left(\frac{\partial F}{\partial V}\right)_T = \left(\frac{\partial P}{\partial T}\right)_V.$$

From (4)

$$\left(\frac{\partial P}{\partial T}\right)_V = -\left(\frac{\partial P}{\partial V}\right)_T \left(\frac{\partial V}{\partial T}\right)_P = \frac{\alpha}{\kappa T}.$$

From (5)

$$C_P - C_V = T\left[\left(\frac{\partial S}{\partial T}\right)_P - \left(\frac{\partial S}{\partial T}\right)_V\right] = T\left[\left(\frac{\partial S}{\partial V}\right)_T \left(\frac{\partial V}{\partial T}\right)_P\right] \quad \text{(Trick 5)}$$

$$= T\left[\left(\frac{\partial P}{\partial T}\right)_V \left(\frac{\partial V}{\partial T}\right)_P\right] = -T\left[\left(\frac{\partial P}{\partial V}\right)_T \left(\frac{\partial V}{\partial T}\right)_P^2\right] \quad \text{(Trick 4)}$$

$$= TV\alpha^2/\kappa_T$$

From (6)

$$\frac{(\partial S/\partial T)_P}{(\partial S/\partial T)_V} = \frac{\partial(S,P)}{\partial(T,P)}\frac{\partial(T,V)}{\partial(S,V)} = \frac{\partial(S,P)}{\partial(S,V)}\frac{\partial(T,V)}{\partial(T,P)} = \frac{(\partial P/\partial V)_S}{(\partial P/\partial V)_T} = \frac{\kappa_T}{\kappa_S}$$

The two results

$$C_P - C_V = \frac{TV\alpha^2}{\kappa_T} \quad \text{and} \quad \frac{C_P}{C_V} = \frac{\kappa_T}{\kappa_S} \tag{4.17}$$

are interesting not only from a calculational point of view; only from them do we see that C_P is greater than C_V even if the substance contracts on heating (e.g., water below 4 °C). Quite generally, the difference and the quotient of such "almost equal" derivatives can be calculated by trick 5 and trick 6, respectively.

Such relations serve to economise effort in actual measurement or in proving the internal consistency of the results of measurement; moreover, by this compact technique one can collect examination points with relative ease.

4.2.3   Alternatives to the Canonical Probability Distribution

The basic axiom (4.1) is called the canonical probability distribution or the canonical ensemble. In many cases it is appropriate to use other assumptions, which in large systems lead to the same results but are easier to calculate.

The canonical case $\rho_n \sim \exp(-\beta E_n)$ corresponds to fixed temperature T, fixed volume V and fixed number of particles N. Therefore $F = -kT \ln Z$ with the partition function $Z = \sum_n \exp(-\beta E_n)$; this free energy F is minimal for fixed T, V and N. Physically the canonical ensemble corresponds, for example, to a container full of water in a bath-tub: V and N are fixed, but through the thin walls heat is exchanged with the water in the bath: T is fixed (Fig. 4.1).

As in the Legendre transformation, we can also work instead with fixed T and fixed μ, which is called the *macrocanonical* ensemble. Here the litre of water in the bath-tub is only in a conceptual volume without real walls, and exchanges not only

Fig. 4.1 Schematic representation of canonical, macrocanonical and microcanonical distribution (from *top* to *bottom*). A litre of water is in a large reservoir of water and heat. *Simple lines* allow heat, but not particles, to pass through; *broken lines* are only imaginary, and *double lines* allow neither heat nor particles through. "Solid" here means fixed in time heat but also particles with its environment. The probability ρ_n of attaining a state Ψ_n with energy E_n and particle number N_n is now

$$\rho_n = \frac{\exp[-\beta(E_n - \mu N_n)]}{Y} \tag{4.18}$$

with the macrocanonical ensemble partition function $Y = \sum_n \exp[-\beta(E_n - \mu N_n)]$ and $J = F - \mu N = -kT \ln Y$. In this macrocanonical ensemble N fluctuates as well as E, whereas T, V, and μ are fixed.

In contrast, we can hold N and E fixed, and then μ and T are allowed to fluctuate:

$$\rho_n \sim \delta(E_n - \langle E \rangle)\, \delta(N_n - \langle N \rangle).$$

This microcanonical ensemble was until a short time ago without much practical use. In recent years the assumption has become important for computer simulations, as the Ising program of electrodynamics, which we used here (Sect. 4.1.1) for the justification of the canonical ensemble, is an approximation for the microcanonical ensemble and works with constant energy and constant number of particles.

Much more important is the macrocanonical ensemble, which we shall need already for the theory of ideal (quantum) gases. If Z_N is the canonical partition function $\sum_n \exp(-\beta E_n)$ at fixed N, then

$$Y = \sum_n Z_N e^{\beta \mu N}. \tag{4.19}$$

From this it follows that

$$\langle N \rangle = \frac{\partial(\ln Y)}{\partial(\beta \mu)} \tag{4.20}$$

analogous to $\langle E \rangle = -\partial(\ln Z)/\partial \beta$ in the canonical ensemble.

In many computer simulations one also works with fixed pressure instead of fixed volume, for which we have already come across the Legendre transformation. Ignoring exceptional cases, all these ensembles are equivalent to the canonical one, i.e. they give the same results. This is easy to understand: if a glass of gin holds 10^{24} molecules then the deviations ΔN about the mean value $\langle N \rangle$ are not critical: after all, what is 10^{12} more or less? You know that even from your Stock Market wealth.

4.2.4 Efficiency and the Carnot Cycle

One of the most impressive results of this 19th century thermodynamics is being able to estimate the efficiency of machines, without making any specific assumptions about the working material. How can one understand steam engines, without knowing something about water?

Fig. 4.2 Pressure as a
function of volume
(schematic) in the four
phases of a Carnot engine
(Sadi Carnot 1796–1828)

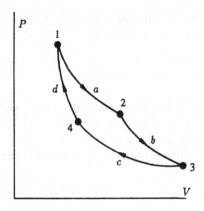

A power unit, whether in a steam locomotive or in an electric power station, converts heat into mechanical work. How mechanical work moves a car (or electrons via the Lorentz force), or whether the heat comes from coal, oil or atomic fission, is for this purpose (but not otherwise) irrelevant. Since an appreciable part of the heat is lost in the cooling water, the efficiency η (η is the ratio of the mechanical work extracted to the quantity of heat supplied) is far below 100%.

A Carnot engine is an ideal power unit, i.e. a cyclic working machine without friction, which is continually in thermal equilibrium. Figure 4.2 shows schematically how the pressure P depends on the volume V in the regular compressions and expansions in the cylinder. We distinguish four phases, each with its transfer of a quantity of heat

$$Q = \int T \mathrm{d}S :$$

(a) isothermal expansion from 1 to 2 (heating)

  $T_1 = T_2, \quad Q = T_1(S_2 - S_1)$

(b) adiabatic expansion from 2 to 3 (insulated)

  $T_2 > T_3, \quad Q = 0, \ S_3 = S_2$

(c) isothermal compression from 3 to 4 (cooling)

  $T_3 = T_4, \quad Q = T_3(S_4 - S_3)$

(d) adiabatic compression from 4 to 1 (insulated)

  $T_4 < T_1, \quad Q = 0, \ S_1 = S_4$

The mechanical work is $A = \int P \mathrm{d}V$, an integral the direct calculation of which assumes knowledge of the properties of the material. The quantity of heat $Q = T \Delta S$ is convenient to work out, since here the temperature is constant during the transfer of heat (steps a and c). According to our assumptions the machine operates cyclically, and so the energy E of the working material at the end of a cycle should be the same as at the beginning:

$$0 = \oint dE = \oint T dS - \oint P dV = T_1(S_2 - S_1) + T_3(S_4 - S_3) - A,$$
$$A = (T_1 - T_3)(S_2 - S_1).$$

In step a the quantity $Q = T_1(S_2 - S_1)$ is invested as heat; accordingly the efficiency $\eta = A/Q$ is equal to the ratio of the temperature difference to the larger of the two temperatures:

$$\eta = \frac{T_1 - T_3}{T_1}. \tag{4.21}$$

The heat not converted into mechanical work goes to the cooling water in step c.

From the theoretical standpoint it is remarkable that the absolute temperature T_1 crops up here: with steam engines one can already establish where the absolute zero lies. In practice power units operate with η between 30 and 40%. The energy consumption of private households depends primarily on heating of air or water (one saves no energy by leaving the electric razor in its case and having instead a wet shave with hot water). One does therefore raise the economic efficiency appreciably by heating dwellings with the cooling water from power units. With nuclear power generators directly on the doorstep, heat utilisation would have a "radiant" future.

Another method of bypassing equation (4.21) is the heat pump, of which the refrigerator is the best known example: using an electric motor we drive the above cyclic process in the reverse direction, and now enquire into the ratio Q/A of the heat Q transferred thereby to the mechanical work A supplied by the motor. From (4.21) it follows that $Q/A = T_1/(T_1 - T_3)$, hence an efficiency far above 100%. In particular you should apply a heat pump to the heating of your swimming-pool, as then T_1 (for sporting folk 18 °C) lies only a little above the temperature T_3 of the soil, from which the heat is extracted. One can, of course, also try to install a refrigerator between the water and the soil.

4.2.5 Phase Equilibrium and the Clausius-Clapeyron Equation

The remaining topics in this section deal with the boundary zone between physics and chemistry, known as physical chemistry (or better as "chemical physics"). It concerns liquids, vapours and binary mixtures in thermodynamic equilibrium. In this section we consider the phase diagram and the gradient of the vapour pressure curve.

Fig. 4.3 Schematic phase
diagram of a simple
substance. P_c and T_c are the
critical pressure and
temperature, respectively,
svp = gas-liquid saturated
vapour pressure curve

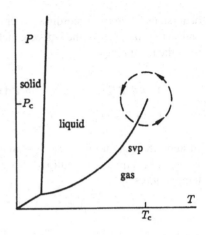

When two "phases", e.g., liquid and vapour of a substance, exist in equilibrium
with each other, and heat, volume and particles are exchanged between them, then
temperature, pressure and chemical potential of the two phases coincide. We prove
this for the temperature: the total energy E_{tot} is a minimum at equilibrium for fixed
S_{tot}, V_{tot}, N_{tot}. Accordingly if a little of the entropy is transferred from the vapour
into the liquid ($dS_{vap} = -dS_{liq}$), the first derivative of the total energy must vanish:

$$0 = (\partial E_{tot}/\partial S_{vap})_{VN} = (\partial E_{liq}/\partial S_{vap})_{VN} + (\partial E_{vap}/\partial S_{vap})_{VN}$$
$$= -(\partial E_{liq}/\partial S_{liq})_{VN} + (\partial E_{vap}\partial S_{vap})_{VN} = -T_{liq} + T_{vap},$$

as stated. This corresponds, moreover, with daily experience, that different states
with thermal contact exchange heat until they have the same temperature. The same
is true for the pressure; only for the chemical potential do we lack sensory feeling.

Figure 4.3 shows schematically the phase diagram of a normal substance with its
two vapour pressure curves, where the vapour is in equilibrium with the liquid and
with the solid, respectively. In reality the pressure rises much more rapidly, e.g., as
four and a half powers of ten with water between the triple point (0 °C) and the critical
point (374 °C). The two vapour pressure curves and the separation curve between
liquid and solid meet at the triple point; the *saturated vapour pressure curve* (svp)
for the transition from gas to liquid ends at the *critical point* $T = T_c$, $P = P_c$.

We shall call the gas a vapour if it is in equilibrium with its liquid, hence lying
on the saturated vapour pressure curve. When we raise the temperature higher and
higher on this curve, the density of the liquid becomes smaller and smaller, and
that of the vapour greater and greater: at the critical point $T = T_c$, $P = P_c$ the two
densities meet at the critical density (0.315 g/cm^3 for water). For $T > T_c$ it is no
longer possible to observe liquid and vapour coexisting: the substance now has only
one homogeneous phase. If one alters P and T so that one passes from the vapour side
of the vapour pressure curve to the liquid side, above T_c (dashed circle in Fig. 4.3),
then the density varies continuously along this path from the vapour value to the

liquid density, without encountering a density discontinuity or two phases. There is accordingly no qualitative difference between liquid and gas; we can distinguish whether a *fluid* is a liquid or gas only when it shows an equilibrium of two phases with $T < T_c$: that with the higher density is then called liquid. Research in 1989 suggests (J. Kertesz, Physica A 161, 58; J.S. Wang, Physica A 161, 249) that along the dashed line of Fig. 4.3 a sharp transition between liquid and gas occurs only in the droplet numbers.

Such a phase transition, during which the difference between two phases continually diminishes, ends at the *critical point* and is called a phase transition of the second order; the phase transition with heating along the vapour pressure curve at $T = T_c$ is one. (The separation curve between the liquid and the solid state does not end in a critical point, since a crystalline solid differs qualitatively from a nonperiodic liquid: a phase transition of the first order.)

This initially surprising fact was discovered in the middle of the 19th century for CO_2, whose T_c lies only a little above room temperature. Because of the high critical pressure of CO_2 it is less dangerous to demonstrate similar phenomena by a separation of binary liquid mixtures. Van der Waals put forward the first theory for this in his doctoral thesis in 1873, the van der Waals equation to be considered later. For air (N_2, O_2) T_c lies at about -150 and $-120\,°C$, respectively; however high a pressure air is subjected to, no liquid droplets will form at room temperature.

On the other hand, if one gradually raises the pressure of a gas at fixed $T < T_c$, then the gas liquefies discontinuously in equilibrium, as soon as the vapour pressure curve is crossed. Such a phase transition with discontinuous change is called a phase transition of the first order. In the weather report one speaks of 100% humidity when the density of water vapour in the air just corresponds to the density on the saturated vapour pressure curve. In reality one needs somewhat higher gas pressure in order to achieve the condensation of vapour into liquid; then first of all very small droplets are formed, in which the energy of the surface tension hinders growth. The presence of condensation nuclei (sea salt, dust raised by rain dances; silver iodide scattered from aeroplanes) encourages the formation of clouds and rain droplets as soon as the humidity is very close to 100%. In completely "nucleus free" air, water vapour condenses only when the humidity is several hundred per cent, in agreement with the nucleation theory 1[1] (about 1930). We can bring about similar nucleation effects in the chemistry laboratory by shaking a test-tube containing a liquid held over a Bunsen burner. Cloud chambers and bubble chambers in high energy physics make use of the fact that electric charges produced by energetic particles can serve as nuclei for the gas-liquid phase change. The photographic process is also, perhaps, a nucleation phenomenon in a phase change of the first order.

In nearly all phase transitions of the first order a latent heat Q occurs, i.e. a quantity of heat Q is needed (or liberated in the reverse change) when one phase changes into another at constant temperature and constant pressure. For example, more heat is needed to boil water at $100\,°C$, than to heat it from $0°$ to $100\,°C$. If one holds one's hand in the steam jet from a boiling kettle, the skin is scalded, not because of the

---

[1] For a conference see Int. J. Mod. Phys. C 3, 773–1164 (1992).

100 °C, but because of the latent heat Q liberated by the water vapour condensing on the skin. We now calculate a relation between Q and the gradient $P' = (\partial P/\partial T)_{svp}$ of the saturated vapour pressure curve.

Using the abbreviations $q = Q/N$, $s = S/N$, $v = V/N$ and $\mu = G/N$ we have; using Table 4.1 and Trick 5:

$$\left(\frac{\partial G}{\partial T}\right)_{svp} = \left(\frac{\partial G}{\partial T}\right)_{P} + \left(\frac{\partial G}{\partial P}\right)_{T}\left(\frac{\partial P}{\partial T}\right)_{svp} = -S + V P'$$

$$-s + v P' = \left(\frac{\partial \mu}{\partial T}\right)_{svp} = \left(\frac{\partial \mu}{\partial T}\right)_{liq} = \left(\frac{\partial \mu}{\partial T}\right)_{vap},$$

since along the saturated vapour pressure curve μ is the same for liquid and vapour. Therefore subtracting $-s + vP'$ for the vapour from the corresponding expression for the liquid, we get zero on the saturated vapour pressure curve:

$$0 = -(s_{liq} - s_{vap}) + (v_{liq} - v_{vap})P'$$
$$= q/T + (v_{liq} - v_{vap})P',$$

from which follows the *Clausius-Clapeyron* equation:

$$q = T(v_{vap} - v_{liq})P'. \tag{4.22}$$

When T is far below T_c the vapour volume per molecule is far greater than the liquid volume: $q = T v_{vap} P'$; one can often assume the vapour to be an ideal gas: $Pv = kT$, or $q = kT^2 P'/P$. Now if the latent heat of condensation is independent of T (low temperatures), then $P' = (q/kT^2)P$ is solved by

$$P \sim e^{-q/kT}. \tag{4.23}$$

From this it is clear that for low temperatures the *latent heat* of condensation q of a molecule is its binding energy in the liquid: in order to liberate the molecule from the liquid and turn it into a molecule of vapour (which then raises the pressure P) the energy q is used up. In actual fact at room temperature the saturated vapour pressure of water doubles for every 10 °C rise in temperature. This rapid rise in the saturated vapour pressure explains many weather phenomena.

4.2.6   Mass Action Law for Gases

Chemical reactions, such as for a hydrogen-oxygen mixture, $2H_2 + O_2 \leftrightarrow 2H_2O$ can generally be written in the form $\sum_i \nu_i A_i \leftrightarrow 0$, where the ν_i are positive or negative integers (namely, the number of molecules taking part per elementary reaction), and A_i the kind of molecule. (The double-headed arrow instead of the ordinary arrow shows that reactions go in both directions.) From the chemists we borrow the notation $[A_i] = N_i/V$ for the concentration; we eschew their other concentration units. If T and V are fixed, then for a fixed total number N of all atoms the individual concentrations $[A_i]$ are determined so that the free energy F is a minimum:

$$0 = dF = \sum_i \left(\frac{\partial F}{\partial N_i}\right)_{TV} dN_i = \sum_i \mu_i dN_i$$

and hence

$$\sum_i \mu_i \nu_i = 0. \tag{4.24}$$

For classical ideal gases we have

$$\mu = kT \ln(N/V) + \text{const.}, \tag{4.25}$$

since

$$\left(\frac{\partial P}{\partial \mu}\right)_{TV} = -\frac{\partial^2 J}{\partial \mu \partial V} = \left(\frac{\partial N}{\partial V}\right)_{TP} = \frac{N}{V}, \quad \text{hence}$$

$$\frac{\partial \mu}{\partial P} = \frac{V}{N} = \frac{kT}{P} \quad \text{or}$$

$$\mu = kT \ln P + \text{Const.}(T) = kT \ln(N/V) + \text{const.}(T).$$

Equations (4.24) and (4.25) together give

$$\sum_i (\ln[A_i] + c_i)\nu_i = 0 \quad \text{or} \quad \sum_i \nu_i \ln[A_i] = C,$$

$$\text{product } \prod_i [A_i]^{\nu_i} = \text{constant } (T, P, \ldots). \tag{4.26}$$

In the hydrogen-oxygen-water reaction the concentration ratio $[H_2]^2[O_2]/[H_2O]^2$ is therefore constant. This mass action law can be generalised to dissociated solutions. When, for example, H_2O dissociates into H^+ and OH^-, then $[H^+][OH^-]$ is constant: if H^+ outweighs OH^- we have an acid, whose pH value is measured by $\log_{10}[H^+]$.

4.2.7 The Laws of Henry, Raoult and van't Hoff

Aqueous solutions of C_2H_5OH form a research field several thousand years old. Concentrated solutions, such as Scotch, are best reserved for professorial study: only weak solutions (e.g., beer) are treated in this section. Suppose, then, that a solute 1 is dissolved in a solvent 2 (e.g., water) with low concentration $c = N_1/(N_1 + N_2) \ll 1$.

We calculate the vapour pressure P_1 and P_2 of the two kinds of molecule over the liquid solution.

The more alcohol there is in the beer, the more alcohol vapour P_1 there is in the air. Quantitatively this is the *Henry* law:

$$P_1 \sim c + \ldots \tag{4.27}$$

for small c. Much more surprising is *Raoult's* law:

$$P_2(c) = P_2(c = 0)(1 - c + \cdots) \tag{4.28}$$

for the vapour pressure of the solvent. Accordingly, if one pours one part per thousand of salt into boiling water, the water ceases to boil, not just because the salt was cold, but rather because it lowers the vapour pressure by one part in a thousand. That the influence of the dissolved substance is proportional to c is understandable; but why is the proportionality factor exactly equal to unity?

The proof rests on the generalised equation (4.16) for the liquid phase: $E - TS + PV - \mu_1 N_1 - \mu_2 N_2 = 0$. The differential is therefore also zero: $-SdT + VdP - N_1d\mu_1 - N_2d\mu_2 = 0$. At fixed (room) temperature and fixed (atmospheric) pressure $dT = dP = 0$ and hence $N_1d\mu_1 + N_2d\mu_2 = 0$. Divided through by $N_1 + N_2$ this gives, still exactly,

$$cd\mu_1 + (1 - c)d\mu_2 = 0$$

for the liquid. In equilibrium each μ is the same for liquid and vapour, and if the vapour is approximated by (4.25) we get for the two vapour pressures

$$c\frac{d(\log P_1)}{dc} + (1 - c)\frac{d(\log P_2)}{dc} = 0. \tag{4.29}$$

So far it is not assumed that $c \ll 1$; by integration one can therefore calculate the vapour pressure of the dissolved substance (to within a factor), if one knows the vapour pressure of the solvent as a function of c. In this way one knew how the vapour pressure of sulphuric acid decreased with dilution in water, long before the vapour pressure of pure sulphuric acid at room temperature was measured in 1975.

For $c \ll 1$ we substitute Henry's law $P_1 \sim c$:

$$\frac{d(\log P_2)}{dc} = -\frac{1}{(1 - c)}$$

or $P_2 \sim (1 - c)$. The proportionality factor must be $P_2(c = 0)$, i.e. the vapour pressure of the pure solvent, from which (4.28) follows.

This lowering of the saturated water vapour pressure raises the boiling point when one adds salt. Correspondingly the freezing point of salt water is lowered. When one scatters salt on the roads in winter one can perhaps melt the ice, corrode the car bodies and salinate the environment. Or instead, one can fly to warmer countries (to important physics conferences, of course).

When one leaves shrivelled prunes in water, one can make them fat and smooth again, or even bring them to bursting-point. This is caused by osmotic pressure. One observes it at semi-permeable membranes, which are layers such as cellophane, and also many biological cell membranes. They let water molecules through, but not more complex dissolved molecules. To the left and right of this semi-permeable membrane, therefore, μ_2 is the same, but μ_1 is not. By the same method as in the Raoult law the change in μ_2 is given by (4.29) as kTc for small c, for then the forces between the N_1 dissolved molecules are zero, as in the ideal gas. From $\partial P_2/\partial \mu_2 = N_2/V$ it follows that there is a pressure difference $P_{\mathrm{osm}} - kTcN_2/V \approx kTN_1/V$, and so

$$P_{\mathrm{osm}} V = N_1 kT + \cdots \tag{4.30}$$

the law of *van't Hoff*.

The molecules of the dissolved substance therefore exert a pressure on the semi-permeable membrane, as if they were molecules of a classical ideal gas. Accordingly one should not drink distilled water, and casualties need at least a physiological solution of salt in their veins.

4.2.8 Joule-Thomson Effect

There arc various methods of causing a sharp drop in the temperature of air. One method is to force it through a throttle valve. Such a valve (known to genuine the-oreticians also as a cotton-wool swab) allows a pressure difference to be maintained without the airflow performing any appreciable mechanical work at the valve. Schematically this looks as in Fig. 4.4.

When the left-hand piston sweeps out a volume dV_1 to the right, it performs the work $P_1 dV_1$. At the right-hand piston the work performed is $P_2 dV_2$. We pump so that the pressure remains constant; moreover, we neglect the heat generated by friction. From energy conservation it then follows that

Fig. 4.4 Schematic layout of an experiment on the Joule-Thomson effect. Air is forced by the piston on the *left* against the valve ×, passes through this and drives the *right*-hand piston through the cylinder. The enthalpy is constant throughout

$$E_1 + \int P_1 dV_1 = (E + PV)_1 = (E + PV)_2 = E_2 + \int P_2 dV_2,$$

i.e. the enthalpy $H = E + PV$ is constant. The question then is: How does the temperature vary in the expansion of a gas when the enthalpy H remains constant?

The calculation of $(\partial T/\partial P)_H$ uses a few of our tricks:

$$(\partial T/\partial P)_H = -(\partial T/\partial H)_P (\partial H/\partial P)_T = -(\partial H/\partial P)_T/(\partial H/\partial T)_P$$
$$= -\frac{(\partial H/\partial P)_S + (\partial H/\partial S)_P (\partial S/\partial P)_T}{(\partial H/\partial S)_P (\partial S/\partial T)_P} = [T(\partial V/\partial T)_P - V]/C_P.$$

Cooling occurs only if $T(\partial V/\partial T)_P > V$. With a classical ideal gas, $PV = NkT$, $\partial V/\partial T = Nk/P = V/T$, and so the whole effect vanishes: $(\partial T/\partial P)_H = 0$. Without forces between the molecules there is no Joule-Thomson effect. The air therefore has first to be cooled to below the "inversion temperature", below which $(\partial T/\partial P)_H$ becomes positive because of the attractive forces between the molecules (cf. virial expansion, next section).

4.3 Statistical Mechanics of Ideal and Real Systems

Up to now in heat theory we have studied general relationships between macroscopic quantities such as the difference between C_P and C_V, based upon mathematical laws. Now we shall work out such quantities directly from microscopic models and, for example, derive the ideal gas laws. Moreover, we shall consider not only exact solutions without interactions ("ideal" gases, etc.), but also approximations for systems with interactions (e.g., "real" gases). In the ideal case it suffices to consider an individual particle or an individual state, since there is no interaction with other particles or states. This simple addition no longer applies in real systems.

4.3.1 Fermi and Bose Distributions

In an ideal gas of force-free molecules how great is the mean number $\langle n_Q \rangle$, of molecules which occupy a certain quantum state, characterised by the set of quantum numbers Q? The Pauli principle of quantum mechanics prevents two or more Fermi particles occupying the same quantum state; for Fermions, therefore, $0 \leq \langle n_Q \rangle \leq 1$, whereas for Bosons $\langle n_Q \rangle$ could become arbitrarily large (and does so with Bose-Einstein condensation). With the canonical, and even more so with the microcanonical ensemble, we now have a problem: if the Pauli principle prevents a particle fitting into one quantum state, so that it must be put into another state, then the different quantum states are no longer statistically independent. It is more convenient to work with the macrocanonical ensemble, where it is not the number of particles, but the chemical potential μ, which is constant: $\rho \sim \exp[-\beta(E - \mu N)]$. Now we need to calculate only a single state (quantum number Q), since the different states have become statistically independent: surplus particles are simply ignored.

The probability of finding n particles in the state with quantum number Q is thus proportional to $\exp[-\beta(\varepsilon - \mu)n]$, since now $n = N$ the particle number and $E = \varepsilon n$ is the energy; ε is thus the energy of a single particle, e.g., $\varepsilon = \hbar^2 Q^2/2m$. With the abbreviation $x = \exp[-\beta(\varepsilon - \mu)]$ the probability is therefore proportional to x^n, with the proportionality factor $1/\sum_n x^n$. In this sum n runs from 0 to ∞ for Bosons and, because of the Pauli principle, from 0 to 1 for Fermions. For Bosons the sum is therefore $= 1/(1 - x)$ ("geometric series"), for Fermions it is $1 + x$. For small x there is accordingly no difference between Bose and Fermi particles, since even for the Bosons the quantum state is hardly ever multiply occupied.

The mean value $\langle n_Q \rangle$ is therefore $\langle n_Q \rangle = \sum_n n x^n / \sum_n x^n$, which gives $x/(1 + x)$ for Fermions. For Bosons the following trick is often useful:

$$\sum_{n=0}^{\infty} n x^n = x \left(\frac{d}{dx}\right) \sum_n x^n = x \left(\frac{d}{dx}\right) \frac{1}{(1 - x)} = \frac{x}{(1 - x)^2},$$

so that $\langle n_Q \rangle = x/(1 - x)$. Fermi and Bose statistics therefore differ only by a sign:

$$\langle n_Q \rangle = \frac{1}{(e^{\beta(\varepsilon - \mu)} \mp 1)} \quad \text{Fermi+,} \quad \text{Bose} - . \tag{4.31}$$

For small x, hence for $\beta(\varepsilon - \mu) \gg 1$, we can replace ± 1 here by zero (classical *Maxwell distribution*): $\langle n_Q \rangle = \exp[\beta(\mu - \varepsilon)]$. If therefore we have particles without forces, $\varepsilon = mv^2/2$, then $\langle n_Q \rangle \sim \exp(-\beta\varepsilon)$ only so long as quantum effects are negligible, $x \ll 1$. The Pauli principle changes the Maxwell distribution well known from classical ideal gases. Figure 4.5 compares the three curves.

The total particle number N and the total energy E are given by the sum over all quantum numbers Q : $N = \sum_Q \langle n_Q \rangle$ and $E = \sum_Q \varepsilon(Q) \langle n_Q \rangle$. Normally the quantum

Fig. 4.5 Comparison of Fermi, Bose and Maxwell distributions, indicated by F, B and M, respectively

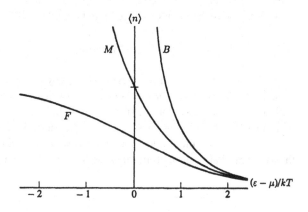

number here is the wave vector Q. In a cube of side length L, L should be an integral multiple of the wavelength $2\pi/Q$, so that $Q_x = (2\pi/L) * m_x$ with $m_x = 0, \pm 1, \pm 2,$ etc. The sum over the integers m_x therefore corresponds to an integral over the x-component Q_x, multiplied by $L/2\pi$, since $dQ_x = (2\pi/L)dm_x$. The triple sum over all three components is therefore given by the calculation rule

$$\sum_Q f(Q) = \left(\frac{L}{2\pi}\right)^3 \int d^3 Q f(Q) \tag{4.32}$$

for arbitrary functions f of the wave vector Q. Of course, one can replace L^3 by the volume V. With $\varepsilon = \hbar^2 Q^2/2m$ we have

$$\sum_Q e^{-\beta\varepsilon} = V/\lambda^3 \tag{4.33}$$

with the "thermal de Broglie wavelength"

$$\lambda = \hbar\sqrt{2\pi/mkT} = \frac{h}{\sqrt{2\pi mkT}}. \tag{4.34}$$

To within factors such as 2π, λ corresponds to the quantum mechanical wavelength $2\pi/Q$ with that momentum $\hbar Q$ for which $\hbar^2 Q^2/2m$ corresponds to the thermal energy kT. Thus, in short: in the classical ideal gas the particles typically have the wavelength λ. If one knows (4.33) one can save much time and effort. Using these results and methods we now discuss various ideal gases of point-like particles in the three limiting cases $\beta\mu \to -\infty$, $\beta\mu \to \infty$, and $\beta\mu \to 0$.

4.3.2 Classical Limiting Case $\beta\mu \to -\infty$

If $\exp(\beta\mu)$ is very small, $\exp[\beta(\varepsilon - \mu)]$ is very large, since the energy $\varepsilon = p^2/2m$ is never negative. Then in (4.31) ± 1 can be neglected:

$$\langle n_Q \rangle = e^{\beta\mu} e^{-\beta\varepsilon(Q)}. \tag{4.35}$$

This Maxwell distribution no longer shows any difference between Fermions and Bosons, and thus corresponds to classical physics. It gives the number of particles with a certain wave vector Q. If one wants the number of all particles with a certain modulus of Q, then $\exp(-\beta\varepsilon)$ has to be multiplied by a factor proportional to $4\pi Q^2 \sim \varepsilon$ because integration is then necessary over all directions. With this factor the Maxwell distribution at first increases, then falls again at large velocities.

The total particle number is

$$N = \sum_Q \langle n_Q \rangle = e^{\beta\mu} \sum_Q e^{-\beta\varepsilon} = e^{\beta\mu} V/\lambda^3,$$

hence

$$\mu = kT \ln(N\lambda^3/V) \tag{4.36}$$

consistent with (4.25), or $\exp(\beta\mu) = (\lambda/a)^3$, with a as the mean particle separation distance ($a^3 = V/N$). This classical limiting case is therefore valid only when $\lambda \ll a$, wavelength smaller than separation distance. For air $a = 30\,\text{Å}$ and $\lambda = 0.2\,\text{A}$, so that the air on earth (certainly not that on Jupiter) can be treated approximately as a classical ideal gas, a happy accident of great significance for school teaching. With metal electrons the converse applies: $\lambda = 30\,\text{Å}$ and $a = 3\,\text{Å}$, so that the conductivity of copper cannot be understood without quanta.

From (4.20) it follows that

$$\frac{\partial(\ln Y)}{\partial(\beta\mu)} = N = \exp(\beta\mu)V/\lambda^3$$

and so by integration

$$\ln Y = \exp(\beta\mu)V/\lambda^3 = N.$$

As we noticed after (4.18), $-kT \ln Y = F - \mu N$ and therefore $-kTN = E - TS - \mu N = -PV$ [using (4.16)]. Hence we obtain the classical ideal gas law

$$PV = NkT, \tag{4.37}$$

which is of course well known. This shows that our definition of the temperature from $\exp(-E/kT)$ is equivalent to the definition from this gas law.

We know $-kTN = F - \mu N$, hence $F = NkT[\ln(N\lambda^3/V) - 1]$ and so

$$S = -\left(\frac{\partial F}{\partial T}\right)_{VN} = Nk\left[\ln(V/N\lambda^3) + \frac{5}{2}\right], \tag{4.38}$$

the Sackur-Tetrode law (1911). (Remark: In differentiating $\ln(V/N\lambda^3)$ with respect to T one does not need to know all the factors, but only that the expression is proportional to $T^{3/2}$; then the derivative is $3/2T$. This trick should be remembered.) Since $E = F + TS$ we have

$$E = \frac{3}{2}NkT, \tag{4.39}$$

hence the other well-known gas law for point-like molecules: the energy per degree of freedom is $kT/2$.

Accordingly we see that the Planck action quantum, which is concealed in the de Broglie wavelength λ, certainly occurs in μ and S, but not in P and E. Without quantum mechanics one can understand the classical ideal gas only partially: as soon as states have to be counted, as in entropy, one needs quantum effects.

Even without quantum mechanics, we can nevertheless draw several conclusions from the entropy:

$$S/Nk = \ln(\text{const.}_1 V T^{3/2}/N)$$
$$= \ln(\text{const.}_2 V E^{3/2}/N^{5/2}) = \ln(\text{const.}_3 T^{5/2}/P) \qquad (4.40)$$

because of (4.37–4.39). For example, in adiabatic expansion, S/N constant, the pressure varies in proportion to $T^{5/2}$; we have

$$C_V = T \left(\frac{\partial S}{\partial T} \right)_{VN} = \frac{3}{2} Nk \quad \text{and} \quad C_P = T \left(\frac{\partial S}{\partial T} \right)_{PN} = \frac{5}{2} Nk,$$

in accordance with the teaching of heat theory in typical school physics.

A further example is the theory of mixtures. Let one litre of argon gas be introduced into the left-hand side of a vessel with an internal dividing partition, and one litre of neon gas into the right-hand side. Now let the dividing partition be removed, and after a little while argon atoms and neon atoms are uniformly mixed in the left- and right-hand sides. This irreversible process has achieved a greater degree of disorder, how great is the increase in entropy? The total entropy is the sum of the entropies of the argon and the neon. Before the removal of the dividing wall the entropy of the argon was

$$S_{Ar} = Nk \ln(\text{const.}_{Ar} V T^{3/2}/N)$$

and that of the neon

$$S_{Ne} = Nk \ln(\text{const.}_{Ne} V T^{3/2}/N).$$

Afterwards, we have

$$S_{Ar} = Nk \ln(\text{const.}_{Ar} 2V T^{3/2}/N) \quad \text{and}$$
$$S_{Ne} = Nk \ln(\text{const.}_{Ne} 2V T^{3/2}/N),$$

because the volume V has doubled. The change in $S_{Ar} + S_{Ne}$ is therefore the mixing entropy:

$$\Delta S = 2Nk \ln 2. \qquad (4.41)$$

Of course, it makes less sense to mix only two particles: heat theory deals with $N \to \infty$.

4.3.3 Classical Equidistribution Law

This law, suggested by (4.39), states that:

> In the classical limiting case every canonical variable (generalised position and momentum) entering quadratically into the Hamilton function (energy) has the mean thermal energy $kT/2$, or in short:
>
> $$\text{energy per degree of freedom} = kT/2 \tag{4.42}$$

For example, the Hamilton function in three dimensions may be:

$$H = \frac{p^2}{2m} + Kr^2 + \frac{L^2}{2\Theta} + \frac{(E^2 + B^2)}{8\pi},$$

whence

$$E = \frac{3kT}{2} + \frac{3kT}{2} + kT + 2kT$$

per molecule. In the angular momentum L we assume here that rotation is possible only about the x- and y-axes (the moment of inertia about the z-axis is so small that $\hbar^2/\Theta \gg kT$, so that rotation about this axis is not stimulated; $\hbar^2/\Theta$ is the smallest possible rotational energy). In electromagnetic waves with a certain wave vector the E- and B-fields must be perpendicular to the wave vector, there are accordingly only two, not three, polarisation directions. In contrast to this, for phonons which arise from $p^2/2m + Kr^2$ there are three directions and consequently a total thermal energy of $3kT$ per particle: Dulong and Petit's law for the specific heat of solids.

One can even prove this equidistribution law; we restrict ourselves to the kinetic energy in one dimension: $H = p^2/2m$. The thermal mean value $2E$ of p^2/m is

$$\begin{aligned}
\langle p^2/m \rangle &= \langle p\partial H/\partial p \rangle \\
&= \frac{\int dp \int dx\, p(\partial H/\partial p)\exp(-\beta H)}{\int dp \int dx \exp(-\beta H)} \\
&= -kT\frac{\int dp \int dx\, p(\partial[\exp(-\beta H)]/\partial p)}{\int dp \int dx \exp(-\beta H)} \\
&= +kT\frac{\int dp \int dx\, \partial p/\partial p \exp(-\beta H)}{\int dp \int dx \exp(-\beta H)} \\
&= +kT,
\end{aligned}$$

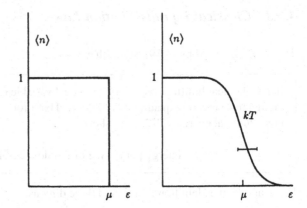

Fig. 4.6 Mean number $\langle n \rangle = \langle n_Q \rangle$ of Fermions in a state characterised by the quantum number Q (e.g., a wave vector). For $T = 0$ on the *left*, for T small $(kT/\mu = 0.1)$ on the *right*

Since we here use position and momentum as separate integration variables, as if there was no quantum mechanical uncertainty, the equidistribution law is valid only in the classical case, without quantum effects.

4.3.4 Ideal Fermi-Gas at Low Temperatures $\beta\mu \rightarrow +\infty$

At low temperatures the classical equidistribution law is no longer valid, the energy is smaller, and one speaks of the freezing of the degrees of freedom. Since in the Fermi-Bose distribution of (4.31) the denominator cannot be zero, and since $\varepsilon = p^2/2m$ varies between 0 and ∞, we must have $\mu \geq 0$ for Fermi-gas and $\mu \leq 0$ for Bose-gas. At very low temperatures the Fermi distribution $\langle n_Q \rangle$ appears as shown in Fig. 4.6: at $T = 0\langle n_Q \rangle$ has a sharp Fermi edge, when $T > 0$ this is smeared over an energy interval of width kT. For $T = 0$, as the total number of particles climbs from zero, first of all the quantum state with the lowest energy is filled, then because of the Pauli principle that with the next lowest energy, and so on. The sharp Fermi edge therefore symbolises the interaction of energy minimisation and the Pauli principle. One then also calls μ the *Fermi energy* ε_F and defines T_F, p_F and Q_F by $\mu = \varepsilon_F = kT_F = p_F^2/2m = \hbar^2 Q_F^2/2m$.

For metal electrons μ/k is of the order of 10^4 K, so that at room temperature $\beta\mu \gg 1$: sharp edge. When the electrons in a solid have quantum states at this Fermi edge, they can be moved by a small electric field, and one has a metal. If, however, there are no electron eigenenergies at the Fermi edge, the electrons with smaller energy cannot change their momentum distribution because of the Pauli principle, and one has an insulator.

For $T = 0$ the Fermi energy μ can be calculated particularly easily:

$$N = \sum_Q \langle n_Q \rangle = V(2\pi)^{-3} \int d^3 Q = V(2\pi)^{-3}(4\pi/3)Q_F^3 = (V/6\pi^2)Q_F^3,$$

or

$$Q_F = (6\pi^2)^{1/3}/a, \quad a^3 = V/N, \quad \mu = \hbar^2 Q_F^2/2m. \tag{4.43}$$

Similarly one calculates the energy $E = \sum_Q \varepsilon(Q)\langle n_Q \rangle$. Division gives

$$\frac{E}{N} = \frac{3}{5}\mu, \quad PV = \frac{2}{3}E. \tag{4.44}$$

The energy per particle is therefore 60% of the maximum energy, and the pressure is the same as in the classical ideal gas. (With N and E experts also take into account a factor $2S + 1$ from the spin S; it cancels out again in E/N.)

For the specific heat C_V we need the spread of the Fermi edge: for small but finite temperatures a fraction kT/μ of all the particles lies in the smeared edge region; each of these particles has a mean energy raised by $\approx kT$ compared with the case $T = 0$. Accordingly the energy E is raised by an amount $\Delta E \approx N(kT/\mu)kT \sim T^2$ compared with $T = 0$, and we have

$$C_V = \left(\frac{\partial E}{\partial T}\right)_V \sim T \tag{4.45}$$

for low temperatures: C_V goes to zero. For metal electrons this law has been well confirmed experimentally, for ^{3}He it is more difficult.

4.3.5 Ideal Bose-Gas at Low Temperatures $\beta\mu \to 0$

The small difference between Fermi- and Bose-gas in the sign of the ± 1 in (4.31) has quite crucial consequences for the particle number N. We have with the abbreviation $z = \varepsilon\mu$ (for spin $= 0$)

$$N = \sum_Q \langle n_Q \rangle = \frac{V}{(2\pi)^3}\int d^3Q \, 1/(e^{\beta(\varepsilon-\mu)} - 1)$$

$$= \frac{V}{2\pi^2}\int dQ \, Q^2/(e^{\beta(\varepsilon-\mu)} - 1)$$

$$= (2V/\lambda^3\sqrt{\pi})\int dz\sqrt{2}/(e^{z-\beta\mu} - 1),$$

where $\beta\mu$ can never be positive (division by zero forbidden).

We calculate this integral by a quite primitive program (BOSE), which finally prints out $N\lambda^3/V$; initial input is $\beta\mu$. The more strongly negative $\beta\mu$ is, the smaller is the integral; it is largest when $\beta\mu = 0$, where the exact calculation gives $s = N\lambda^3/V = 2.61$.

PROGRAM BOSE

```
 10  bm=-1.0
 20  s=0.0
 30  for iz=1 to 100
 40  z=0.1*iz
 50  s=s+sqr(z)/(exp(z-bm)-1)
 60  print s
 70  next iz
 80  s=s*0.2/sqr(3.14159)
 90  print s
100  end
```

What do we do now, if the particle density N/V exceeds this limiting value $2.61\lambda^{-3}$? Does the vessel explode? The latter is indeed improbable, since neither forces nor a Pauli principle act between the Bosons. In reality *Bose-Einstein condensation* (1925) takes place: the superfluous particles, which do not fit into the above integral, form a kind of residue, in momentum space, of course. We therefore again look at

$$\langle n_Q \rangle = \frac{1}{e^{\beta(\varepsilon-\mu)} - 1}.$$

If $\mu = 0$ and ε is very small, then

$$\langle n_Q \rangle = \frac{1}{e^{\beta\varepsilon} - 1} \approx \frac{kT}{\varepsilon} = \frac{1}{z}$$

is very large. As $\varepsilon \to 0$ this number of particles diverges, and such divergence is not correctly approximated by the above integral: equation (4.32) is no longer valid if the function to be summed has a very sharp peak. We therefore simply replace $N = \sum_Q \langle n_Q \rangle$ by $N = N_0 + \sum_Q \langle n_Q \rangle$ with the number N_0 of the particles with zero energy:

$$N = N_0 + \frac{2V}{\lambda^3 \sqrt{\pi}} \int dz \frac{\sqrt{z}}{e^{z-\beta\mu} - 1}.$$

This equation thus describes a new kind of phase transition, one of the few which can be calculated exactly in three dimensions:

$$\text{If} \quad \frac{N}{V} < 2.61\lambda^{-3}, \quad \text{then} \quad \frac{N_0}{V} = 0 \quad \text{and} \quad \mu < 0.$$

$$\text{If} \quad \frac{N}{V} > 2.61\lambda^{-3}, \quad \text{then} \quad \frac{N_0}{V} > 0 \quad \text{and} \quad \mu = 0.$$

The N_0 particles in the ground state therefore provide a finite percentage of all the particles if $N/V \gg \lambda^{-3}$, whereas in the normal case $N/V \ll \lambda^{-3}$ the number of particles with $\varepsilon = 0$ is a small finite number, whose fractional contribution to all the

N particles tends to zero as $N \to \infty$. We therefore learn quite fortuitously that phase transitions are sharp only when $N \to \infty$. The limiting condition $N/V = 2.61\lambda^{-3}$ gives the transition temperature

$$T_0 = (2\pi\hbar^2/mk)(N/2.61\,V)^{2/3}, \tag{4.46}$$

which to within dimensionless factors agrees with the Fermi temperature T_F, although the physics behind it is quite different. For $T < T_0$ there is a condensate N_0, for $T > T_0$ there is not. (Anybody interested in critical phenomena and scaling laws may calculate Bose-Einstein condensation in d dimensions, with not only integer values of d between 2 and 4, see Chap. 5. We find here the universality class of the n-vector model with $n \to \infty$.)

For $T < T_0$, therefore, $\mu = 0$ and the fraction $1 - N_0/N$ of the particles, which do *not* belong to the condensate, is proportional to $\lambda^{-3} \sim T^{/2}$ because of the above integral; when $T = T_0$ this becomes unity:

$$\begin{aligned} N_0/N &= 1 - (T/T_0)^{3/2} \quad &(T < T_0) \\ &= 0 \quad &(T > T_0). \end{aligned} \tag{4.47}$$

The ideal Bose-gas below T_0 therefore consists of a normal liquid with thermal motion, and the condensate whose particles are all at rest. This condensate has no energy, no entropy, no specific heat and no friction (friction in gases arises from collisions between particles in thermal motion). Accordingly, if one could allow such a Bose-gas to flow through a fine capillary, the condensate would flow through, but the normal component would not. Such phenomena have been observed experimentally in "superfluid" ^{4}He (also known as helium II) below the lambda-temperature of 2.2 K. In superfluid helium there is also frictionless vortex motion, corresponding to our theories in hydrodynamics. The above formula for T_0 gives about 3 K for ^{4}He. Of course, the lambda-temperature and T_0 do not coincide exactly, as the helium atoms exert forces upon each other. Therefore the connection between condensate (ideal Bose gas) on the one hand and the superfluid component (real ^{4}He) on the other hand is loosened.

Metal electrons are Fermi particles and therefore undergo no Bose-Einstein condensation. However, two electrons in an elastic solid can attract each other and form a so-called "Cooper pair". These Cooper pairs are Bosons and therefore become superfluid at low temperatures. Since they carry electric charges one speaks of superconductivity, which is explained by the BCS theory (Bardeen, Cooper and Schrieffer 1957). Until 1986 the superconducting condensation temperatures lay below 25 K. Then Bednorz and Müller of IBM Zurich made the experimental breakthrough to higher, more easily attainable temperatures, and soon afterwards 95 K was confirmed; in March 1987 this sensational development led to a conference of physicists, which the New York Times dubbed the "Woodstock" of the physicists.[2] Its practical applications were not as large as was hoped. Of less importance technically, but theoretically

---

[2]For a conference see Physica C 185–189 (1991).

just as interesting, is the fact that the Fermi particles of ^{3}He also form pairs and, as has been known since 1972, can be superfluid,[3] although only at 10^{-3} K.

4.3.6 Vibrations

What contribution to the specific heat is made by vibrations of all kinds, i.e. the phonons, photons and other quasi-particles of the harmonic oscillator in Sect. 3.2.6? At high temperatures with $kT \gg \hbar\omega$ the equidistribution law must hold, but what happens at low temperatures? The mean thermal energy E_ω of an oscillator is

$$E_\omega = \hbar\omega \left(\langle n \rangle + \frac{1}{2} \right), \quad \langle n \rangle = \frac{1}{e^{\beta\hbar\omega} - 1}, \tag{4.48}$$

which one can also derive from the partition function $\sum_n \exp[-\beta\hbar\omega(n + \frac{1}{2})]$; then one obtains

$$F_\omega = \frac{\hbar\omega}{2} + kT \ln(1 - e^{-\beta\hbar\omega}).$$

The chemical potential is zero since, in contrast to real Bosons, there is no constant particle number for the Bose quasi-particles. In a medium with various vibration frequencies we sum over all quantum numbers Q for the total energy $E = \sum_Q \hbar\omega(Q)(\langle n_\omega \rangle + \frac{1}{2})$.

We are chiefly interested in waves with wave vector Q and a frequency $\omega(Q)$ proportional to Q^b, e.g., $b = 1$ for phonons and photons. With $\omega^{-1}d\omega = bQ^{-1}dQ$ we then have in d dimensions, ignoring the constant zero-point energy (from $\hbar\omega/2$):

$$E = \sum_Q \hbar\omega\langle n(Q) \rangle \sim \int d^d Q \omega \langle n(Q) \rangle \sim \int dQ\, Q^{d-1} \omega \langle n \rangle \sim \int d\omega\, Q^d \langle n \rangle$$

$$\sim \int d\omega\, \omega^{d/b}/(e^{\beta\omega} - 1) \sim T^{1+d/b} \int dy\, y^{d/b}(e^y - 1) \sim T^{1+d/b},$$

since the integral over $y = \beta\hbar\omega$ converges and gives a contribution to the proportionality factor. The specific heat $\partial E/\partial T$ is then

$$C_V \sim T^{d/b}. \tag{4.49}$$

At low temperatures, therefore, C_V becomes very small, in contrast to the equidistribution law, according to which it must remain constant: freezing of the degrees of freedom. The above calculation is normally valid only for low temperatures, since

---

[3]J.C. Wheatley, page 1 in: Progress in Low Temperature Physics, vol. VIIa, edited by C.J. Gorter (North Holland, Amsterdam 1978).

$\omega \sim Q^b$ usually only for small ω and since in the above integral the main contribution comes from y near 1, and hence from $\hbar\omega$ near kT.

Experimentally confirmed applications of this result are:

phonons $b = 1$	$d = 3$	$C \sim T^3$	DEBYE law
photons $b = 1$	$d = 3$	$C \sim T^3$	STEFAN-BOLTZMANN law
magnons $b = 1$	$d = 3$	$C \sim T^3$	for antiferromagnetism
magnons $b = 2$	$d = 3$	$C \sim T^{3/2}$	for ferromagnetism
ripplons $b = 3/2$	$d = 2$	$C \sim T^{4/3}$	ATKINS law

Here magnons are the quantised magnetisation waves, in which the three-dimensional vector M rotates its direction periodically; ripplons are surface waves (with a two-dimensional wave vector) in the superfluid helium, which thus make a contribution $\sim T^{7/3}$ to the surface tension ($C_V = dE/dT \sim T^{4/3}$).

For photons the Stefan-Boltzmann law holds not only at low temperatures, since $\omega = cQ$ even for large Q: $E/V = (\pi^2/15)(kT)^4/(\hbar c)^3$ including all the factors. The fact that the main contribution to the energy comes from frequencies ω near $kT/\hbar$ is called the Wien displacement law. In this way the temperature of the surface of the sun is known to be about 6,000 K, and that of the universal (background) radiation 3 K. (If in summer it is 30 °C in the shade, how hot is it in the sun? Answer: In equilibrium it is 6,000°.) According to the equidistribution law, $E_Q = 2kT$ comes from each wave vector Q, hence an infinitely high energy. This nonsensical result for the "black body radiation" was the starting point for the quantum theory (Max Planck 1900).

4.3.7 Virial Expansion of Real Gases

This expansion, whose name remains unexplained here, is a perturbation theory (Taylor expansion) in terms of the strength of the forces between the molecules, in order to correct the ideal gas equation $PV = NkT$:

$$PV/N\,kT = 1 + B(T)N/V + C(T)(N/V)^2 + \cdots . \tag{4.50}$$

We neglect quantum effects and obtain after laborious algebra (or simple copying):

$$2B = \int \left(1 - e^{-\beta U}\right) d^3r \tag{4.51}$$

with the potential $U = U(r)$ for two particles at a distance r. We assume that one can find a radius r_c with $U \gg kT$ for $r < r_c$ and $U \ll kT$ for $r > r_c$; then we have $1 - e^{-\beta U} = 1$ in the first case and $= \beta U$ in the second case:

$$2B = 4\pi r_c^3/3 + \int_{r>r_c} \beta U(r) d^3r = 2b - a/kT. \qquad (4.52)$$

For spheres with radius $r_c/2$, b is four times the volume; the integral for a is usually negative, since $U < 0$ for medium and large distances. In fact measurements of the second virial coefficient $B = B(T)$ show that it is constant and positive at high temperatures, but with cooling it becomes first smaller and then negative.

4.3.8 Van der Waals' Equation

Better than this exact virial expansion is the van der Waals approximation, since it leads also to the phase transition to a liquid. First of all we rewrite

$$N kT = \frac{PV}{(1 + BN/V + \cdots)}$$

as

$$N kT = PV(1 - BN/V) = P(V - BN) \approx P(V - bN) = P V_{\text{eff}}$$

with the effective volume V-four times the volume of all the particles approximated by spheres. This takes account of the collisions, and we still need a correction W for the attraction:

$$F = -NkT(1 + \ln(V_{\text{eff}}/N\lambda^3)) + W.$$

The first term corresponds to the free energy F mentioned before (4.38), and W must allow for the attraction.

We take $W/N = \int U(r)N/V d^3r$ as a crude approximation which assumes that the probability of the presence of other particles at distance r from a given molecule is not influenced by this molecule and is therefore given by N/V. Comparison with (4.52) shows that

$$\frac{W}{N} = -a\frac{N}{V},$$

and hence

$$F = -N kT(1 + \ln[(V - bN)/N\lambda^3] - aN^2/V);$$

Fig. 4.7 Representation of isotherms from (4.53). The shaded areas are equal and determine the equilibrium for $T < T_c$

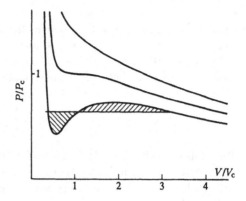

with $P = -\partial F/\partial V$ we have

$$N\,kT = (V - bN)(P + aN^2/V^2). \tag{4.53}$$

In reality one does not calculate b and a as they are derived here, but chooses a and b so that this van der Waals equation gives the best possible agreement with experiment.

This approximation produces not only gaseous but also liquid behaviour, and leads to the continuity between the two phases described in Sect. 4.2.5. For temperatures below a critical temperature T_c the isotherms $P = P(V)$, at constant T, are no longer monotonically falling, but show a minimum and a maximum, as shown in Fig. 4.7. A portion of this curve corresponds to supersaturated vapour and supercooled liquid, before equilibrium is established by nucleation. The equilibrium volumes of liquid and vapour on the vapour pressure curve are obtained by the *Maxwell construction*: a horizontal line is taken through the $P - V$ diagram so that it cuts off areas of equal size below the maximum and above the minimum, as shown in Fig. 4.7.

The critical temperature is therefore determined so that the isotherm $P = P$ $(V, T = T_c)$ has a point of inflection with a horizontal tangent at the critical point $P = P_c$, $V = V_c$. At this point we have

$$\left(\frac{\partial P}{\partial V}\right)_T = \left(\frac{\partial^2 P}{\partial V^2}\right)_T = 0$$

and therefore

$$P - P_c \sim (V_c - V)^3 + \dots.$$

This condition leads to

$$V_c = 3bN, \quad P_c = \frac{a}{27b^2}, \quad kT_c = \frac{8a}{27b}, \quad \frac{P_c V_c}{N\,kT_c} = \frac{3}{8}. \tag{4.54}$$

Experimentally $P_c V_c / N k T_c$ is nearer to 0.3 and is affected by electric dipole moments on the molecules as well as by quantum effects for light atoms. Moreover, the exponent δ in $P - P_c \sim |V - V_c|^\delta$ is experimentally not three, but nearly five.

With $P^* = P/P_c$, $V^* = V/V_c$, $T^* = T/T_c$, these equations (4.53) and (4.54) combine to give the dimensionless universal equation of state:

$$\frac{8T^*}{3} = \left(V^* - \frac{1}{3} \right) \left(P^* + \frac{3}{V^{*2}} \right). \tag{4.55}$$

The relationship between the dimensionless variables $T*$, $V*$ and $P*$ summarises the properties of all substances. If one wishes to know the vaporisability of gold, one only needs to bring water to the boil and then convert to the corresponding T, V and P. [Even the quark plasma has perhaps a critical temperature: Phys. Lett. B 241, 567 (1990)]. Unfortunately in practice it is not accurate, and the universal equation of state has to be replaced by the much less powerful principle of universality, which is valid only close to the critical point (next section).

4.3.9   Magnetism of Localised Spins

With the van der Waals equation we have avoided the mathematical derivation of the critical point, as we have that more conveniently for the ferromagnetic Curie point in the framework of the mean field theory. We again work with the Ising model from Sect. 2.2.2 on electrodynamics, where the spins $S_i = \pm 1$ of a lattice are pointed either upwards or downwards. A pair of neighbouring spins S_i and S_j make the contribution $-J_{ij} S_i S_j$ to the Hamilton operator (energy). In an external magnetic field B there is also the magnetic energy $-B S_i$ for each spin.

We set the magnetic dipole moment equal to unity; strictly speaking we should write $-B \mu_B S_i$. We see also that "$-$ field $\times$ moment" is strictly speaking a Legendre transform, since this term contributes to the energy in a fixed field. In Sect. 4.1.1 on classical thermodynamics, however, the energy had the extensive quantities such as S, V, N as natural variables, and therefore here the magnetisation and the electric polarisation. Accordingly if we now assume probability proportional to $\exp(-\beta E)$ and therefore add $-B \sum_i S_i$ to E, then this is a macrocanonical probability analogous to $\exp[-\beta(E - \mu N)]$, with BM for μN. In the literature, however, this significant distinction is seldom made. H is even often written instead of B; in reality one means the magnetic field acting on an individual spin, and hence the sum of the external field and all the dipole interactions of the other spins.

First we need a counterpart for the ideal gas, and this is provided by spins without interactions: $J_{ij} = 0$. It suffices then to consider a single spin, which has, according to orientation, the "energy" $\pm B$. The probability of the spin pointing upwards in the direction of the magnetic field is

$$w_+ = \frac{e^{\beta B}}{e^{\beta B} + e^{-\beta B}},$$

and pointing downwards is

$$w_- = \frac{e^{\beta B}}{e^{\beta B} + e^{-\beta B}}$$

and the mean value $m = \langle S_i \rangle$ is

$$w_+ - w_- = \frac{e^{\beta B} - e^{-\beta B}}{e^{\beta B} + e^{-\beta B}} = \tanh(\beta B).$$

The total magnetisation M of the N spins (per cm^3) is

$$M = N \tanh(\beta B). \tag{4.56}$$

For the initial susceptibility $\chi = (\partial M / \partial B)_{B=0}$ we have the Curie law:

$$\chi = \frac{N}{kT}, \tag{4.57}$$

since $\tanh (B/kT) \approx B/kT$ for small B.

If we now include the spin-spin interaction J_{ij}, the total energy (Hamilton operator) is

$$H_{\text{Ising}} = -\sum_i \sum_j J_{ij} S_i S_j - B \sum_i S_i.$$

We approximate S_j by its thermal mean value $m = \langle S_j \rangle$ (still to be worked out) and find

$$H_{\text{Ising}} = -\sum_i \left(\sum_j J_{ij} m \right) S_i - B \sum_i S_i = -B_{\text{eff}} \sum_i S_i,$$

where the effective field $B_{\text{eff}} = B + m \sum_j J_{ij}$. This sum $\sum_j J_{ij} = kT_c$ is the same in a solid for all lattice points i. That the T_c so defined is the Curie temperature will now be established. The interaction energy H_{Ising} is accordingly reduced to the case treated above of the spins without interaction in an effective magnetic field B_{eff}. According to (4.56) we have

$$M/N = m = \tanh(\beta B_{\text{eff}}) = \tanh(\beta B + mT_c/T) \tag{4.58}$$

with the above definition of T_c. This mean field theory is therefore analogous to the van der Waals equation, since here too the influence of a particle on its neighbours

Fig. 4.8 Magnetic field
(*left*) and free energy
$F = F' + BM$ (*right*) from
(4.61)

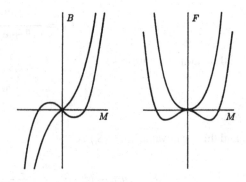

is not fully taken into account: there we made the approximation $W/N \approx -aN/V$, and here $S_i S_j \approx S_i m$.

For $B = 0$ the equation $m = \tanh(mT_c/T)$ has only one solution $m = 0$ for $T > T_c$, but two more solutions $\pm m_0$ for $T < T_c$. The spontaneous magnetisation $M_0 = Nm_0$ is therefore different from zero below T_c, so that T_c is the Curie temperature. Above the Curie temperature we have, for $\beta B \ll 1$: $\tanh(x) \approx x$ and hence $\beta B = (1 - T_c/T)m$. The initial susceptibility Nm/B is therefore

$$\chi = N/k(T - T_c) \quad \text{(Curie-Weiss)}. \tag{4.59}$$

For magnetic fields, calculations near the critical point are easiest if m and B are small and T is close to T_c. Because $\tanh(x) \approx x - x^3/3$ it follows that $m = mT/T_c + \beta B - m^3/3 + \cdots$, or

$$\beta B = \frac{T - T_c}{T}m + \frac{m^3}{3}, \quad M = Nm. \tag{4.60}$$

Figure 4.8 shows the isotherms in the $B - M$ diagram, clearly analogous to the $P - V$ diagram of Fig. 4.7. Now it is already clear on the basis of symmetry that the areas below the maximum and above the minimum must be the same. The three different types of solution for m when $B = 0$ become clearer from the free energy $F'(B) = F - BM$ with $B = \partial F/\partial M$:

$$F'/NkT = C + (1 - T_c/T)m^2/2 + m^4/12 - Bm$$
$$= C + \text{const.}_1(T - T_c)m^2 + \text{const.}_2 m^4 - Bm. \tag{4.61}$$

The last line has the form of the Landau Ansatz (1937) for critical phenomena and holds more generally than the mean field theory; e.g., it is valid in five dimensions.

Above the Curie temperature there is only a minimum of the free energy F, at $M = 0$: paramagnetism. Below T_c this stable minimum becomes an unstable maximum, and in addition there occur two minima at $\pm M_0$: ferromagnetism with spontaneous magnetisation M_0. Equation (4.60) gives this spontaneous magnetisation as $m_0 =$

$\sqrt{3}[(T_c - T)/T]^{1/2}$, whereas just at $T = T_c$ the magnetic field $\sim m^3$. Below T_c, F has a contribution $(1 - T_c/T)m_0^2 + m_0^4/12$, which gives a jump in the temperature profile of the specific heat.

Unfortunately these exponents are just as imprecise as those of the van der Waals equation. Experimentally m_0 varies roughly like $(T_c - T)^{1/3}$ in three dimensions and like $(T_c - T)^{1/8}$ in two dimensions; and at $T = T_c$ the magnetic field is proportional roughly to m^5 in three dimensions and m^{15} in two. The scaling theory of 1965 succeeded in setting these critical exponents in relation to each other, Kenneth G. Wilson's theory of renormalisation (Nobel Prize 1982) explained them.[4]

The similarity between the critical point of gases and liquids and the Curie point of ferromagnets is no accident: we only need to identify $S_i = 1$ with an occupied lattice position and $S_i = -1$ with a free one, then we have a model of the lattice gas: liquid for upward spin, vapour bubble for downward spin. In actual fact all liquids at the critical point appear to have the critical exponent of the three-dimensional Ising model ("universality").

PROGRAM METROPOLIS

```
 10  dim is (1680), w(9)
 20  L=40
 30  t=2.5
 40  L1=L+1
 50  Lp=L*L+L
 60  Lm=Lp+L
 70  for i=1 to Lm
 80  is (i)=1
 90  next i
100  for ie=1 to 9 step 2
103  ex=exp(-2*(ie-5)/t)
106  w(ie)=ex/(1.0+ex)
109  next ie
110  for it=1 to 100
120  m=0
130  for i=L1 to Lp
140  ie=5+is(i)*(is(i-1)+is(i+1)+is(i-L)+is(i+L))
145  if rnd(i)<w(ie) then is(i)=-is(i)
150  m=m+is(i)
160  next i
170  print it,m
180  next it
190  end
```

---

[4]Phase Transitions and Critical Phenomena, edited by C. Domb, M.S. Green, and J.L. Lebowitz, vols. 1–16 (Academic Press, London 1972–1994).

The program METROPOLIS now modifies the Ising program of Sect. 2.2.2 on electrodynamics according to the canonical ensemble: fixed temperature, fluctuating energy. Here one follows the quite general principle of the Monte Carlo simulation of Metropolis et al. (1953). It is named for the roulette tables of Monte Carlo since Las Vegas was less known at that time; the randomness of the roulette result is simulated by the random number generator rnd in the program:

1. Choose a spin.
2. Calculate the energy ΔE of reversal.
3. Calculate a random number z between 0 and 1.
4. Reverse it if $z <$ the probability.
5. Calculate the required quantities if necessary.
6. Choose a new spin and start again.

For the probability we take $\exp(-\beta\Delta E)/(1 + \exp(-\beta\Delta E))$; so the sum of the two probabilities is unity (to flip or not to flip, that is the question). T is input as kT/J, whence, if the nearest neighbour interaction is J, the Curie temperature is given by $J/kT_c = \frac{1}{2}\ln(1 + \sqrt{2}) = 0.44$ or 0.22165 on a square or simple cubic lattice, respectively, instead of the mean field theory values 1/4 and 1/6.

The program adds a 5 to the energy index ie, so that $W(i.e.)$ always has a positive index (as required by many computers); ignoring that fact, $ie = S_i \sum_j S_j$ is half of the reversal energy $\Delta E/2J$. By skillful plotting one can produce beautiful clusters, more or less fractal, and compare them with the experimental results of Fig. 4.9.

Fig. 4.9 Optical investigation of the concentration fluctuations ("clusters") in a liquid mixture of isobutyric acid and water at $T - T_c = 0.001$ K. The picture corresponds to a square of side 0.2 mm. (Perrot, Guenon and Beysens, Saclay 1988). Computer simulations give similar clusters which are fractal near T_c; see Chap. 5

4.3.10 Scaling Theory

How does the scaling theory work, which was mentioned after the Landau Ansatz (4.61) as giving better critical exponents? We rewrite this Landau Ansatz for the normalized free energy

$$f = \frac{F' + B M}{N kT} - C$$

in the more complicated form

$$f_s = f/|T - T_c|^2 = \text{const.}_1 \left(m/|T - T_c|^{1/2}\right)^2 + \text{const.}_2 \left(m/|T - T_c|^{1/2}\right)^4$$
$$= \text{const.}_1 m_s^2 + \text{const.}_2 m_s^4 + \cdots \tag{4.62}$$

with the scaled magnetization $m_s = m/|T - T_c|^{1/2}$ and the scaled free energy $f_s = f|T - T_c|^2$. (If you dislike arbitrary powers of a dimensioned temperature, simply replace $T - T_c$ by the dimensionless difference $(T - T_c)/T$.) We now regard (4.62) as a Taylor expansion in powers of the scaled magnetization m_s for a more complicated function $f_s(m_s)$; there are no odd terms in this expansion, like m_s^3, since the free energy f_s does not change if the magnetization is switches its sign. With this reinterpretation, the expansion (4.62) is valid only for reasonably small m_s. Moreover, since the exponents of this Landau Ansatz disagree with experiment, we generalize our definitions of f_s and m_s to

$$f_s = \frac{f}{|T - T_c|^{2-\alpha}}, \quad m_s = \frac{m}{|T - T_c|^\beta}, \tag{4.63}$$

with critical exponents α and β taken from experiment instead of from theory. Thus

$$f = |T - T_c|^{2-\alpha} f_s(m_s) = |T - T_c|^{2-\alpha} f_s \left(\frac{m}{|T - T_c|^\beta}\right), \tag{4.64}$$

with the general scaling function f_s depending on the scaled variable m_s only. The magnetic field $B = \partial F/\partial M$ then is

$$B = |T - T_c|^{2-\alpha-\beta} b_s(m_s), \tag{4.65}$$

where b_s is the derivative of f_s with respect to m_s. Thus the free energy F and the magnetic field B are self-similar functions of M and $T - T_c$: if near T_c we measure $F(M)$ or $B(M)$ at some temperature, then the same curves are obtained at another temperature provided we rescale the free energy by $|T - T_c|^{2-\alpha}$, the magnetization by $|T - T_c|^\beta$ and the field by $|T - T_c|^{2-\alpha-\beta}$.

These scaling laws (4.62) are today widely believed to be exact in "asymptotia", i.e. in the asymptotic limit $T \to T_c$, $M \to 0$, $B \to 0$. The zero-field susceptibility $\chi = 1/(\partial B/\partial M)$, is, according to (4.65), proportional to $|T - T_c|^{2-\alpha-2\beta}$, whereas right at $T = T_c$, the scaling function $b_s(m_s)$ must for very large arguments vary as $m_s^{(2-\alpha-\beta)/\beta}$ in order that the temperature difference $T - T_c$ cancels out; thus: $B \sim M^{(2-\alpha-\beta)/\beta}$ for small magnetizations M. The exponents γ and δ are defined through $\chi \sim |T - T_c|^{-\gamma}$ and $B \sim M^{\delta}$ in zero field or at T_c, respectively; thus

$$2 - \alpha = \gamma + 2\beta = \beta(\delta + 1). \tag{4.66}$$

The zero-field specific heat is proportional to the second temperature derivative of the free energy and thus varies as $|T - T_c|^{-\alpha}$. In this sense, two of the critical exponents α, β, γ, and δ are sufficient to determine the other exponents, just as in classical thermodynamics we could derive some measurable quantities from other such quantities. But in contrast to thermodynamics, these scaling laws of 1965 are valid only very close to the Curie point. (The fractal dimension d_f to be mentioned in the next chapter is $d/(1 + 1/\delta)$ at the Curie point).

Similar scaling laws have been found for other phase transitions and other properties. For example, the size distribution for the clusters visible, e.g., in Fig. 4.9, is described by two exponents (from which α, β, γ, and δ can be derived) and is studied through "percolation" theory; the behavior near a liquid-gas critical point is analogous to that of the Curie point, provided we identify M with the density difference from the critical point, and B with the chemical potential difference. Then all known critical exponents for the liquid-gas critical point agree with that of the 3-dimensional Ising model, independent of material: approximately $\alpha = 0.11$, $\beta = 0.32$, $\gamma = 1.24$, $\delta = 4.8$, $d_f = 2.5$.

Questions

Section 4.1

1. How many molecules are there in a cubic centimetre of air?
2. How long should we or a computer need to play through all the magnetic configurations of a system of $L \times L \times L$ spin $1/2$ particles, for $L = 1, 2, 3$ and 4?

Section 4.2

3. What does the differential of the energy look like?
4. What is the Legendre transformation and what is it used for?
5. Which quantity is minimal in equilibrium for fixed T, V, N, M and which for fixed T, P, N, B (M = magnetisation, B = field)?
6. How does one define C_V, C_P and χ_M?
7. Which two quantities are related by the Clausius-Clapeyron equation?
8. State van't Hoff's law on osmotic pressure.

Section 4.3

10. What are the Fermi, Bose and Maxwell distributions?
11. How does $S = S(E, V, N)$ look in the classical ideal gas?
12. With what power of T does $C_V(T \to 0)$ vary in the ideal Fermi-gas?
13. What are Fermi energy, Fermi temperature and Fermi momentum?
14. How does $C_V(T \to 0)$ of a vibration ω depend on T?
15. With what power of T does $C_V(T \to 0)$ vary in optic phonons, and in acoustic phonons?
16. With what power of T does $C_V(T \to 0)$ vary in spin waves?
17. What is the virial expansion and what is it used for?
18. What is the "Maxwell construction" in the van der Waals equation?
19. State the law of corresponding states.
20. What is the "equation of state" $M = M(B, T)$ for spins without interactions, and what is it in mean field theory?
21. With what power of M or $V - V_c$ does B or $P - P_c$, respectively, vary at $T = T_c$ (mean field theory, van der Waals equation, reality)?

Problems

Section 4.1

1. A finite system has ten quantum states with $E = 1$ erg, 100 with $E = 2$ erg, and 1000 with $E = 3$ erg. What are the mean energy and entropy, if all states are occupied equally strongly (infinitely high temperature), and what are they at $20\,°C$?
2. Of N interaction-free spins, each should have equal probability of pointing upwards or downwards. What probability is there of having just m spins upwards? Using the Stirling formula $[n! \approx (n/e)^n]$, approximate this probability for $N \to \infty$. How does the width of the resulting Gauss curve depend on N?

Section 4.2

3. Check the following relationships:

   (a) $(\partial T/\partial V)_{SN} = -(\partial P/\partial S)_{VN}$,
   (b) $(\partial T/\partial P)_{SN} = -(\partial V/\partial S)_{PN}$,
   (c) $(\partial P/\partial S)_{TN} = -(\partial V/\partial T)_{PN}$,
   (d) $(\partial V/\partial M)_{TBN} = -(\partial B/\partial P)_{TVN}$,
   (e) $C_V/C_P = \kappa_T/\kappa_S$,
   (f) $\chi_T/\chi_S = C_M/C_B$,
   (g) $C_B - C_M = T(\partial M/\partial T)/\chi_T$.

4. Recast the following expressions in more significant form:

   (a) $(\partial P/\partial N)_{TV}$,
   (b) $(\partial P/\partial N)_{SV}$,
   (c) $(\partial T/\partial N)_{SP}$,
   (d) $(\partial \mu/\partial S)_{TV}$,
   (e) $\partial(P/T)/\partial N)_{EV}$ (think!),
   (f) $(\partial F/\partial V)_{SN}$,
   (g) χ_V/χ_P.

5. Show that $PV = 2E/3$ for the ideal inert gas ($\varepsilon = p^2/2m$) and $PV = E/3$ for the light quantum gas ($\varepsilon = cp$), in which free particles are reflected elastically at the walls and so exert pressure.
   Hint: $\langle p_x v_x \rangle = \langle pv \rangle/3$ in three dimensions.

Section 4.3

6. In the classical limiting case what are the mean kinetic and potential energies of:

   (a) a particle with $\varepsilon = p^{10}/m$ in one dimension,
   (b) an anharmonic oscillator $H = p^2/2m + Kx^{10}$ (1-dimensional),
   (c) an extremely relativistic inert gas ($\varepsilon = cp$) (d-dimensional).

7. Calculate for $T \rightarrow 0$ the specific heat of a solid with one longitudinal and two transverse acoustic phonon branches.

8. From the virial expansion up to the second coefficient B, show that the inversion temperature, at which the Joule-Thomson effect vanishes, is given by the maximum in B/T.

9. Calculate the quantity a in the second virial coefficient $B = b - a/2kT$ for the Lennard-Jones potential $U = 4\varepsilon[(\sigma/r)^{12} - (\sigma/r)^6]$ and $r_c = \sigma$.

10. Express the magnetisation fluctuations with a fixed magnetic field B through the susceptibility χ.

11. Calculate the partition function and the free energy in the magnetic field B for a particle with spin $= 1/2$ without mutual interactions.

12. Calculate the ratio of the initial susceptibility above and below T_c, at the same distance from T_c, using the Landau expansion or mean field theory.

Chapter 5
Fractals in Theoretical Physics

Abstract Random walks and Sierpinski gaskets are used to introduce fractal dimensions, and diffusion limited aggregates finish this "dessert", which is closely related to the critical exponents at the end of the previous chapter.

This chapter is intended to be read as a 'dessert'–a kind of reward for having worked through the four main pillars of theoretical physics presented thus far.[1] No background from the previous chapters is assumed, so the reader who skips the main meal is nonetheless welcome to taste the dessert. We won't be like the conscientious wife who denied her dying husband's last wish—to taste the freshly baked cakes whose odor drifted from her kitchen to his deathbed—with the scolding remark "The cakes are for *after* the funeral!"

Plato sought to explain nature with five regular solids; Newton and Kepler bent Plato's circle to an ellipse; modern science analyzed Plato's shapes into particles and waves, and generalized the curves of Newton and Kepler to relative probabilities—still without a single 'rough edge'. Now, more than two thousand years after Plato, three centuries after Newton, and after strenuous decades of wily insinuation, calculated argument, and stunning demonstration, Benoit Mandelbrot (1924–2010) has established a discovery that ranks with the laws of *regular* motion. Bespeaking the knowledge possessed by every child and every great painter, Mandelbrot has observed, "Clouds are not spheres, mountains are not cones, coastlines are not circles, bark is not smooth, nor does lightning travel in a straight line".

What Mandelbrot has named fractal geometry describes not only the zigzag of Zeus's thunderbolt, or the branching and the varying densities of Pan's forests. It describes as well the Mercurial irregularities of the commodities market, the heretofore unaccountable fits of Poseidon the earthshaker, and a myriad of phenomena in the realm of lesser deities—snowflakes, shale, lava, gels, the rise and fall of rivers, fibrillations of the heart, the surging of electronic noise. Fractal geometry points to a symmetry of pattern within each of the meldings, branchings, and shatterings of nature.

---

[1] For a picture album see E. Guyon and H.E. Stanley, Fractal Forms (Elsevier, Amsterdam 1991).

© Springer-Verlag GmbH Germany 2017   189
D. Stauffer et al., *From Newton to Mandelbrot*, Graduate Texts in Physics,
DOI 10.1007/978-3-662-53685-8_5

A book that preceded by more than half a century Mandelbrot's 1982 classic *The Fractal Geometry of Nature* and was known by every scientist at that time is *On Growth & Form* by W. D'Arcy Thompson (1917). *On Growth & Form* called attention to the fact that a large part of science was based on structures and processes that on a microscopic level are completely random, despite the fact that on the macroscopic level we can perceive *patterns* and *structure*. This classic has become popular again, in large part due to the fact that in the past few years the advent of advanced computing and sophisticated experimental techniques have led to dramatic progress in our understanding of the connection between the structure of a variety of random 'forms' and the fashion in which these forms 'grow'. Not surprisingly, within the scientific community there has been a tremendous upsurge of interest in this opportunity to unify a large number of diverse phenomena, ranging from chemistry and biology to physics and materials science.

5.1   Non-random Fractals

Fractals fall into two categories, *random* (Plate 1) and *non-random* (Plate 2). Fractals in physics belong to the first category, but it is instructive to discuss first a much-studied example of a non-random fractal—the Sierpinski gasket. We simply iterate a *growth rule* much as a child might assemble a castle from building blocks. Our basic unit is a triangular-shaped tile shown in Fig. 5.1a, which we take to be of unit 'mass' ($M = 1$) and of unit edge length ($L = 1$).

The Sierpinski gasket is defined operationally as an 'aggregation process' obtained by a simple iterative process. In stage one, we join three tiles together to create the structure shown in Fig. 5.1b, an object of mass $M = 3$ and edge $L = 2$. The effect of stage one is to produce a unit with a lower density: if we define the density as

$$\rho(L) = M(L)/L^2, \tag{5.1}$$

Fig. 5.1 First few stages in the aggregation rule which is iterated to form a Sierpinski gasket fractal

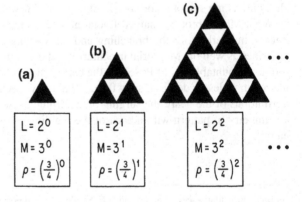

then the density decreases from unity to 3/4 as a result of stage one.

Now simply iterate—i.e., repeat this growth rule over and over *ad infinitum*. Thus in stage two, join together—as in Fig. 5.1c—three of the $\rho = 3/4$ structures constructed in stage one, thereby building an object with $\rho = (3/4)^2$. In stage three, join three objects identical to those constructed in stage two. Continue until you run out of tiles (if you are a physicist) or until the structure is infinite (if you are a mathematician!). The result after stage four—with 81 black tiles and 175 white tiles (Fig. 5.2a) may be seen to this day in floor mosaics of the church in Anagni, Italy, which was built in the year 1104 (Plate 2). Thus although the Sierpinski gasket fractal is named after a 20th century Polish mathematician, it was known some eight centuries earlier to every churchgoer of this village!

The citizens of Anagni did not have double-logarithmic graph paper in the 12th century. If they had had such a marvelous invention, then they might have plotted the dependence of ρ on L. They would find Fig. 5.2b, which displays two striking features:

1. $\rho(L)$ decreases monotonically with L, without limit, so that by iterating sufficiently we can achieve an object of *as low a density as we wish*, and
2. $\rho(L)$ decreases with L in a *predictable* fashion, namely a simple power law.

Power laws have the generic form $y = Ax^\alpha$ and, as such, have two parameters, the 'amplitude' A and the exponent α. The amplitude is not of intrinsic interest, since it depends on the choice we make for the definitions of M and L. The exponent, on the other hand, depends on the process itself—i.e., on the 'rule' that we follow when

Fig. 5.2 a Sierpinski gasket fractal after four stages of iteration, **b** a log-log plot of ρ, the fraction of space covered by *black tiles*, as a function of L, the linear size of the object we iterate. In short, different rules give different exponents. In the present example, $\rho(L) = L^\alpha$ so the amplitude is unity. The exponent is given by the slope of Fig. 5.2b,

$$\alpha = \text{slope} = \frac{\log 1 - \log(3/4)}{\log 1 - \log 2} = \frac{\log 3}{\log 2} - 2 . \tag{5.2}$$

Finally we arc ready to define the fractal dimension d_f, through the equation

$$M(L) = AL^{d_f} . \tag{5.3}$$

If we substitute (5.3) into (5.1), we find

$$\rho(L) = AL^{d_f - 2} . \tag{5.4}$$

Comparing (5.2) and (5.4), we conclude that the Sierpinski gasket is indeed a fractal object with fractal dimension

$$d_f = \log 3 / \log 2 = 1.58... \tag{5.5}$$

Classical (Euclidean) geometry deals with regular forms having a dimension the same as that of the embedding space. For example, a line has $d = 1$, and a square $d = 2$. We say that the Sierpinski gasket has a dimension intermediate between that of a line and a square.

We may generalize the Sierpinski gasket from $d = 2$ to $d = 3$, taking as the basic building block a regular tetrahedron of edge $L = 1$ and mass $M = 1$. Combining *four* such blocks we can build a $L = 2$ tetrahedron with a hole in the center—so that $M = 4$ for $L = 2$, and this construction may be iterated indefinitely to form an object resembling the Great Pyramid after a termite attack. We see that $d_f = 2$, so for this example the fractal dimension is an integer! We offer as an amusing exercise to generalize the Sierpinski gasket structure to an embedding space of *arbitrary* dimension d (yes, we can have exercises during dessert, provided that they are amusing). You should find the result

$$d_f = \log(d + 1) / \log 2 . \tag{5.6}$$

5.2 Random Fractals: The Unbiased Random Walk

Real systems in nature do not resemble the floor of the Anagni church—in fact, no non-random fractals are found in Nature. What is found are objects which themselves are not fractals but which have the remarkable feature that if we form a *statistical average* of some property such as the density, we find a quantity that decreases linearly with length scale when plotted on double logarithmic paper. Such objects are termed *random fractals,* to distinguish them from the non-random *geometric fractals* discussed in the previous section.

Consider the following prototypical problem in statistical mechanics. At time $t = 0$ an ant[2] is parachuted to an arbitrary vertex of an infinite one-dimensional lattice with lattice constant unity: we say $x_{t=0} = 0$. The ant carries an *unbiased* two-sided coin, and a metronome of period one. The dynamics of the ant is governed by the following rule. At each 'tick' of the metronome, it tosses the coin. If the coin is heads, the ant steps to the neighboring vertex on the East [$x_{t=1} = 1$]. If the coin is tails, it steps to the nearest vertex on the West [$x_{t=1} = -1$].

Are there *laws of nature* that govern the position of this drunken ant? At first thought, the response is likely to be "*NO—How can you predict the position of something that is random?*" However, if you have reached this far in a primer on theoretical physics, then you can imagine that there may be laws governing even the motion of random systems.

For the drunken ant described above, the first 'law' concerns $\langle x \rangle_t$, the expectation value of the position of the ant after a time t. In general, the expectation value of *any* quantity A is given by

$$\langle A \rangle = \sum_c A_c P_c , \tag{5.7}$$

where A_c is the value of the quantity A in configuration c, P_c is the probability of configuration c, and the summation is over all configurations. For the example at hand, there are 2 configurations at time $t = 1$ with $P_c = 1/2$, 4 configurations at time $t = 2$ with $P_c = 1/4$. In general, there are 2^t configurations at an arbitrary time t, each with probability $P_c = (1/2)^t$. Thus

$$\langle x \rangle_t = \sum_c x_c P_c = 0 \tag{5.8}$$

for $t = 1$. To prove (5.8) in general, proceed by induction: assume (5.8) holds for time t and show that it holds for time $t + 1$.

For *non-random* systems, it is generally sufficient to predict the position of the system at time t—the analog of $\langle x \rangle_t$. For *random* systems, on the other hand, the information contained in $\langle x \rangle_t$ does not describe the system extensively. For example, we know intuitively that as time progresses, the average of the *square* of the displacement of the ant increases monotonically. The explicit form of this increase is contained in the second 'law' concerning the *mean square displacement*

$$\langle x^2 \rangle_t = t . \tag{5.9}$$

---

[2]The use of the term ant to describe a random walker is used almost universally in the theoretical physics literature—perhaps the earliest reference to this colorful animal is a 1976 paper of de Gennes that succeeded in formulating several general physics problems in terms of the motion of a 'drunken' ant with appropriate rules for motion. Generally speaking, classical mechanics concerns itself with the prediction of the position of a 'sober' ant, given some set of non-random forces acting on it, while statistical mechanics is concerned with the problem of predicting the position of a drunken ant.

Equation (5.9) may also be proved by induction, by demonstrating that (5.9) implies $\langle x^2 \rangle_{t+1} = t + 1$.

Additional information is contained in the expectation values of higher powers of x, such as $\langle x^3 \rangle_t$, $\langle x^4 \rangle_t$, and so forth. By the same symmetry arguments leading to (5.8), we can see that $\langle x^k \rangle_t = 0$ for all *odd* integers k. However $\langle x^k \rangle_t$ is non-zero for *even* integers. Consider, e.g., $\langle x^4 \rangle_t$. We may easily verify that

$$\langle x^4 \rangle_t = 3t^2 - 2t = 3t^2 \left[1 - \frac{2/3}{t} \right]. \tag{5.10}$$

5.3   'A Single Length'

5.3.1   The Concept of a Characteristic Length

Let us compare (5.9) and (5.10). What is the displacement of the randomly walking ant? On the one hand, we might consider identifying this displacement with a length L_2 defined by

$$L_2 = \sqrt{\langle x^2 \rangle} = t^{1/2}. \tag{5.11}$$

On the other hand, it is just as reasonable to identify this displacement with the length L_4 defined by

$$L_4 = \sqrt[4]{\langle x^4 \rangle} = \sqrt[4]{3}\, t^{1/2} \left[1 - \frac{2/3}{t} \right]^{1/4}. \tag{5.12}$$

The important point is that both lengths display an asymptotic dependence on the time. We call the leading exponent (i.e., 1/2) the *scaling exponent*, while the non-leading exponents are termed *corrections-to-scaling*. The reader may verify that the same scaling exponent is found if we consider any length L_k (provided k is even),

$$L_k = \sqrt[k]{\langle x^k \rangle} = A_k\, t^{1/2} \left[1 + B_k t^{-1} + C_k t^{-2} + \cdots + O\left(t^{1-k/2} \right) \right]^{1/k}. \tag{5.13}$$

The subscripts on the amplitudes indicate that these depend on k. Equation (5.13) exemplifies a robust feature of random systems: *regardless of the definition of the characteristic length, the same scaling exponent describes the asymptotic behavior.* We say that all lengths scale as the square root of the time, meaning that whatever length L_k we choose to examine, L_k will *double* whenever the time has increased by a factor of 4. This scaling property is not affected by the fact that the amplitude A_k in (5.13) depends on k, since we do not inquire about the absolute value of the length L_k but only enquire how L_k *changes* when t changes.

5.3.2 Higher Dimensions

Next, we shall show that the identical *scaling laws* hold for dimensions above one. Suppose we replace our one-dimensional linear chain lattice with a two-dimensional square lattice. This entails replacing our ant's coin with a four-sided bone.[3] According to the outcome of the 'bone toss', the ant will step North, East, South, or West. The coordinate of the ant is represented by a two-dimensional vector $r(t)$ with Cartesian components $[x(t), y(t)]$.

The analogs of (5.8) and (5.9) are

$$\langle r \rangle_t = 0 \tag{5.14}$$

and

$$\langle |r|^2 \rangle_t = t . \tag{5.15}$$

We may formally prove (5.14) and (5.15) by induction. Equation (5.15) may also be 'understood' if we note that, on average, for half the metronome ticks the ant steps either to the East or to the West, so from (5.9) the x-displacement should follow the law $\langle x^2 \rangle_t = t/2$. The other half of the time the ant moves North or South, so $\langle y^2 \rangle_t = t/2$. Hence $\langle |r|^2 \rangle_t = \langle x^2 \rangle_t + \langle y^2 \rangle_t = t$.

For the fourth moment, we find

$$\langle |r|^4 \rangle_t = 2t^2 \left[1 - \frac{1/2}{t} \right] . \tag{5.16}$$

Thus the length L_4 defined in (5.12) scales with the *same* scaling exponent for two dimensions as for one dimension; the amplitudes of the leading terms and the 'correction-to-scaling' term are changed, but the asymptotic scaling properties are not affected in passing from $d = 1$ to $d = 2$. A hallmark of modern critical phenomena is that the *exponents* are quite robust but *amplitudes depend more sensitively* on what particular system is being studied.

5.3.3 Additional Lengths that Scale with $\sqrt{t}$

Linear polymers are topologically linear chains of monomers held together by chemical bonds (like a string of beads). Let us make an oversimplified model of such a linear polymer by assuming that the chain of monomers adopts a conformation in three-dimensional space that has the same statistics as the *trail* of the ant. By the trail we mean the object formed if the ant leaves behind a little piece of bread at each

---

[3] Montroll and Shlesinger have written that ancient cave men (and presumably cave women) were fascinated by games of chance and would actually roll four-sided bones to randomly choose one of four possible outcomes.

site visited. After a time t, the ant has left behind t pieces of bread; hence the analog of the time is the number of monomers in the polymer chain. An unrealistic feature of this simple model arises whenever the ant re-visits the same site. Then more than one piece of bread occupies the same site, while two monomers *cannot* occupy the same point of space. In Sect. 5.8, we shall see that statistical properties of a random walk provide a useful upper bound on the properties of real polymers, and that this upper bound becomes the exact value of d_f for space dimensions above a critical dimension d_c.

We can experimentally measure the radius of gyration R_g of this random walk model of a polymer. Moreover, it is a simple exercise to demonstrate that

$$R_g = \frac{1}{\sqrt{6}} R_{EE} , \tag{5.17}$$

where $R_{EE} = \sqrt{\langle |r|^2 \rangle}$ is the Pythagorean distance between the first and last monomer; R_{EE} is called the end-to-end distance of the random walk. Thus we expect that R_g scales as the square root of the number of monomers, just as the lengths L_2 and L_4 of (5.11) and (5.12) scale as the square root of the time.

Thus we find identical scaling properties no matter what definition we choose— the moment L_k of (5.13), the radius of gyration R_g of the trail, or the end-to-end displacement of the entire walk. In this sense, there is only 'one characteristic length'. When such a characteristic length is referred to, generically, it is customary to use the symbol ξ.

5.4   Functional Equations and Scaling: One Variable

We have seen that several different definitions of the characteristic length ξ all scale as $\sqrt{t}$. Equivalently, if $t(\xi)$ is the characteristic time for the ant to 'trace out' a domain of linear dimension ξ, then

$$t \sim \xi^2 . \tag{5.18}$$

More formally, for all positive values of the parameter λ such that the product $\lambda\xi$ is large, $t(\xi)$ is, asymptotically, a *homogeneous function*,

$$t(\lambda^{1/2}\xi) = \lambda t(\xi) . \tag{5.19}$$

Equation (5.19) is called a functional equation since it provides a constraint on the form of the function $t(\xi)$. In contrast, algebraic equations provide constraints on the numerical values of the quantities appearing in them. In fact, (5.18) is the 'solution' of the functional equation (5.19) in the sense that any function $t(\xi)$ satisfying (5.19) also satisfies (5.18)—we say that power laws are the solution to the functional equation (5.19). To see this, we note that if (5.19) holds for all values of the parameter λ, then it holds in particular when $\lambda = 1/\xi$. With this substitution, (5.19) reduces to (5.18).

It is also straightforward to verify that any function $t(\xi)$ obeying (5.18) obeys (5.19). Thus (5.19) implies (5.18) *and conversely*. This connection between power law behavior and a symmetry operation, called *scaling symmetry*, is at the root of the wide range of applicability of fractal concepts in physics.

5.5 Fractal Dimension of the Unbiased Random Walk

Writing (5.18) in the form

$$t \sim \xi^{d_f} \tag{5.20}$$

exhibits the feature that the scaling exponent d_f explicitly reflects the asymptotic dependence of a characteristic 'volume' (the number of points in the trail of the ant) on a characteristic 'length' (R_g, R_{EE}, or L_k). Thus for the random walk, $d_f = 2$, but in general d_f is a kind of dimension. We call d_f the *fractal dimension* of the random walk.

If we write (5.19) in the form

$$t(\lambda\xi) = \lambda^{d_f}t(\xi)\,, \tag{5.21}$$

then we see that d_f plays the role of a scaling exponent governing the *rate* at which we must scale the time if we wish to trace out a walk of greater spatial extent. For example, if we wish a walk whose trail has twice the size, we must wait a time 2^{d_f}. Similarly, if we wish to 'design' a polymer with twice the radius of gyration, we must increase the molecular weight by the factor 2^{d_f}.

It is significant that the fractal dimension d_f of a random walk is 2, *regardless of the dimension of space*. This means that a time exposure of a 'drunken firefly' in three-dimensional space is an object with a well-defined dimension,

$$d_f = 2\,. \tag{5.22}$$

Similarly, a time exposure in a Euclidean space of any dimension d produces an object with the identical value of the fractal dimension, $d_f = 2$.

5.6 Universality Classes and Active Parameters

5.6.1 Biased Random Walk

Next we generalize to the case in which the motion of the ant is still random, but displays a bias favoring one direction over the other. We shall see that the bias has the effect of changing, *discontinuously*, the exponent characterizing the dependence on time of the characteristic length.

Let us place our ant again on a one-dimensional lattice, but now imagine that its coin is *biased*. The probability to be heads is

$$p = \frac{1+\varepsilon}{2} , \tag{5.23}$$

while the probability to be tails is $q = 1 - p = (1 - \varepsilon)/2$. From (5.23) we see that the parameter

$$\varepsilon = 2p - 1 = p - q . \tag{5.24}$$

defined in (5.23) is the difference in probabilities of heads and tails; ε is called the *bias*. We say that such an ant executes a *biased random walk*.

Although the results of the previous section will be recovered only in the case $\varepsilon = 0$, the same general concepts apply. The possible configurations of the biased walk are the same as for the unbiased random walk—i.e., we say that the phase space is the same. The values A_c associated with each configuration (each point in phase space) are also the same. However, instead of being identically $(1/2)^t$ for all configurations, the values of P_c now depend upon the configuration. If events are uncorrelated, then the joint probability is simply the product of the separate probabilities. Hence

$$P_c = p^{h_c}(1 - p)^{t-h_c} , \tag{5.25}$$

where h_c is the number of 'heads' in configuration c.

5.6.2   Scaling of the Characteristic Length

Now the expectation value $\langle x \rangle_t$ is not zero, as it was for the unbiased ant. Rather, we find that (5.8) is replaced by

$$\langle x \rangle_t = (p - q)t = \varepsilon t . \tag{5.26}$$

Thus the bias ε plays the role of the *drift velocity* of the center of mass of the probability cloud of the ant, since the time derivative of $\langle x \rangle_t$ is the analog of a velocity.

Other expectation values are also affected. For example, (5.9) generalizes to

$$\langle x^2 \rangle_t = [(p - q)t]^2 + 4pqt = \varepsilon^2 t^2 + (1 - \varepsilon^2)t . \tag{5.27}$$

If $\varepsilon = p - q = 0$, the results (5.26) and (5.27) reduce to (5.8) and (5.9). We thus recover the unbiased ant, for which the characteristic length ξ scales as $\sqrt{t}$. For any non-zero value of ε, no matter how small, we see from (5.26) and (5.27) that asymptotically

Fig. 5.3 **a** The *discontinuous* change in fractal dimension d_f for the biased random walk as the active parameter $\varepsilon = p - q$ is varied. **b** The *continuous* change in $\langle x^2 \rangle$ as a function of time for a small value of the bias parameter $\varepsilon = p - q = 10^{-3}$. Note the crossover between the apparent fractal dimension $d_f = 2$ for $t \ll t_x$ to the asymptotic fractal dimension $d_f = 1$ for $t \gg t_x$, where $t_x = 1/\varepsilon^2$ is the crossover time

$$L_k = \sqrt[k]{\langle x^k \rangle} \sim t. \qquad (5.28)$$

for $k = 1, 2$ respectively (the general-k result is a bit of an exercise!). Thus we conclude that the ξ scales linearly in time: the fractal dimension of the walk changes *discontinuously* with ε from $d_f = 1$ for all non-zero ε to $d_f = 2$ for $\varepsilon = 0$ (Fig. 5.3).

Systems with the same exponent are said to belong to the same *universality class*. We say that the biased walk belongs to the $d_f = 1$ universality class for all non-zero values of the parameter ε, and that it belongs to the $d_f = 2$ universality class for $\varepsilon = 0$. The term *active parameter* is used to describe a parameter such as ε which changes the universality class of a system.

Here is a paradox! The dependence of d_f on bias ε is a *discontinuous* function of ε, yet the actual motion of the ant cannot differ much as ε changes infinitesimally. To resolve this paradox, consider a specific example of a biased walk with an extremely small value of bias, say $\varepsilon_0 = 10^{-6}$. The r.h.s. of (5.27) has two terms. If only the first term were present, the ant would simply 'drift' to the right with uniform velocity ε. If only the second term were present, the motion of the biased ant would be the same as that of the unbiased ant, except that the width of the probability distribution would be reduced by a factor $(1 - \varepsilon^2)$. To see which term dominates, we express the r.h.s. as $[\varepsilon^2 t + 1]t$. We can now define an important concept, the *crossover time* $t_x = 1/\varepsilon^2$. For $t \ll t_x$ the second term dominates and the ant has the statistics of an *unbiased* random walk; we say that the trail has an *apparent* fractal dimension $d_f = 2$. For $t \gg t_x$, the first term dominates and the ant has the statistics of a *biased* random walk; the trail assumes its *true* or asymptotic fractal dimension $d_f = 1$ (Fig. 5.3b). Note that the crossover time t_x is quite large if the bias is small. If the bias is, say,

0.001, then the ant must walk a million steps before its trail becomes distinguishable from that of an unbiased ant!

Analogous considerations govern the crossover from one universality class to another in thermal critical phenomena, of the sort discussed in Chap. 4. Thus, e.g., if we have a three-dimensional magnet with interactions much weaker in the z direction, then far from the critical point the system displays apparent two-dimensional behavior, while close to the critical point it crosses over to its true asymptotic three-dimensional behavior. Thus we see that the important concepts of universality classes and the phenomenon of crossover between universality classes both have a geometric counterpart in the behavior of the biased random walk in the limit of small bias fields.

5.7 Functional Equations and Scaling: Two Variables

In this section we generalize the concept of a homogeneous function from one to two independent variables. We say a function $f(u, \upsilon)$ is a *generalized homogeneous function* if there exist two numbers a and b (termed scaling powers) such that for all positive values of the parameter λ, $f(u, \upsilon)$ obeys the obvious generalization of (5.19),

$$f(\lambda^a u, \lambda^b \upsilon) = \lambda f(u, \upsilon) .   \tag{5.29}$$

We can see by inspection of (4.64) that the free energy near the critical point obeys a functional equation of the form of (5.29), so generalized homogeneous functions must be important! To get a geometric feeling for such functions and their properties, consider the simple Bernoulli probability $\Pi(x, t)$—the conditional probability that an ant is found at position x at time t given that the ant started at $x = 0$ at $t = 0$. In the *asymptotic* limit of large t, $\Pi(x, t)$ is expressible in closed form (unlike the free energy near the critical point!). The result is the familiar Gaussian probability density

$$\Pi_G(x, t) = \frac{1}{\sqrt{2\pi t}} \exp\left[-\frac{x^2}{2t} \right] ,   \tag{5.30}$$

Note that $\Pi_G(x, t)$ clearly satisfies (5.29), with scaling powers $a = -1$ and $b = -2$,

$$\Pi_G(\lambda^{-1} x, \lambda^{-2} t) = \lambda \Pi_G(x, t) .   \tag{5.31}$$

The predictions of the scaling relations (5.31) are given by the properties of generalized homogeneous functions. Among the most profound and useful of these properties is that of *data collapsing*. If (5.31) holds for all positive λ, then it must hold for the particular choice $\lambda = t^{1/2}$. With this choice, (5.31) becomes

$$\frac{\Pi_G(x, t)}{t^{-1/2}} = \Pi_G\left(\frac{x}{t^{1/2}}, 1 \right) = F(\tilde{x}) ,   \tag{5.32}$$

Fig. 5.4 Schematic illustration of scaling and data collapse as predicted by (5.32) for $\Pi_G(x, t)$, the Gaussian probability density

where we have defined the *scaled variable* $\tilde{x}$ by

$$\tilde{x} = \frac{x}{t^{1/2}} \, . \tag{5.33}$$

Equation (5.32) states that if we 'scale' the probability distribution by dividing it by a power of t, then it becomes a function of a *single* scaled distance variable obtained by dividing x by a different power of t. Instead of data for $\Pi(x, t)$ falling on a family of curves, one for each value of t, data *collapse* onto a single curve given by the *scaling function* $F(\tilde{x})$ (Fig. 5.4). This reduction from a function of n variables to a function of $n - 1$ *scaled* variables is a hallmark of fractals and scaling. The 'surprise' is that the function $F(\tilde{x})$ defined in (5.32) at first sight would seem to be a function of *two* variables, but it is in fact a function of only a single scaled variable $\tilde{x}$.

5.8 Fractals and the Critical Dimension

Thus far we have seen that the study of fractals help us in understanding two developments of modern theoretical physics:

1. The empirical fact that the equation of state simplifies greatly in the vicinity of a critical point, and
2. The empirical fact that diverse systems behave in the identical fashion near their respective critical points—a fact given the rather pretentious name *universality*.

Here we discuss one more simplification that occurs near critical points: above a certain *critical dimension* the mean field theory of Sect. 4.3.9 suffices to determine the critical exponents! This remarkable fact can be understood better using simple geometric concepts.

In Sect. 5.3.3, we introduced a geometric object with the same fractal properties as the trail of a random walk. This object is called a linear polymer, treated in the 'free-flight' approximation in which we can neglect the intersections of the chain with itself. Of course, no two objects can really occupy the same point in space, a fact known at least since the time of Archimedes' famous bathtub experiments.

Fig. 5.5 Schematic illustration of **a** a random walk, and **b** a self-avoiding walk (SAW). Each has taken 6 steps. We show just one of the 4^6 possible 6-step walks—many of these have zero weight for the SAW case. Shown also are schematic log-log plots showing how many steps are needed (the 'mass' M of the trail) for the walk to explore a region of characteristic size ξ, where here ξ is identified with the mean end-to-end distance R_{EE}

Hence the random walk model of a polymer chain cannot suffice to describe a real polymer. Instead, real polymers are modeled by a *self-avoiding walk* (SAW) in which a random walker must obey the 'global' constraint that he cannot intersect his own trail (Fig. 5.5).

A remarkable fact is that in sufficiently high spatial dimensions the SAW has the *identical* fractal dimension as the unbiased random walk, because in sufficiently high dimension the probability of intersection is so low as to be negligible. To see this, we first note that the *co-dimension* $d - d_f$ of the fractal trail is an exponent governing how the fraction of space 'carved out' by the trail decreases with length scale L, since from (5.1) ρ decreases as $\rho(L) \sim M(L)/L^d \sim (1/L)^{d-d_f}$. Now if two fractal sets with dimensions d_f' and d_f'' intersect in a set of dimension $d_\cap$, then the *sum* of the co-dimensions of the two sets is equal to the co-dimension of the intersection set,

$$d - d_\cap = (d - d_f') + (d - d_f'') . \tag{5.34}$$

This general result follows from the fact that a site belongs to the intersection only if it belongs to both fractals: since statistically independent probabilities multiply (p. 131), the fraction of space (with exponent $d - d_\cap$) carved out by *both* fractals is the product of the fractions of space (with exponents $d - d_f'$ and $d - d_f''$) carved out by each.

To apply (5.34) to the trail of a random walk, consider the trail as being two semi-infinite trails—say red and blue—each with random walk statistics. If we substitute $d_f' = d_f'' = 2$ in (5.34), we find that for d equal to a critical dimension $d_c = 4$ the red and blue chains will intersect in a set of zero dimension. Thus for $d > d_c$, the 'classical' random walk suffices to describe the statistical properties of self-avoiding polymers!

The counterpart of this geometric statement is that the simple 'classical' theories presented in Chap. 4 give correct exponents for all dimensions above some critical dimension d_c. Indeed, this is one of the key results of recent years in theoretical physics. In this regard, we now introduce two generalizations of the simple Ising model which appear to be sufficient for describing almost all the universality classes necessary for understanding critical phenomena (Fig. 5.6).

The first generalization of the Ising model is the Q-state Potts model. Each spin ζ_i localized on site i assumes one of Q *discrete orientations* $\zeta_i = 1, 2, \ldots, Q$. If two neighboring spins ζ_i and ζ_j have the same orientation, then they contribute an amount—J to the energy, while if ζ_i and ζ_j are in different orientations, they contribute nothing. Thus the total energy of an entire configuration is

$$\varepsilon(Q) = -J \sum_{\langle ij \rangle} \delta(\zeta_i, \zeta_j), \tag{5.35}$$

where the Kronecker symbol again is defined as

$$\delta(\zeta_i, \zeta_j) = \begin{cases} 1 & \text{if } \zeta_i = \zeta_j \\ 0 & \text{otherwise} . \end{cases} \tag{5.36}$$

Fig. 5.6 Schematic illustration of a 'Metro map' showing how the Ising model has been generalized, first to form a *'North-South'* line (allowing the two Ising spin orientations to become Q discrete orientations—the Potts model), and then to form an 'East-West' line (allowing the two spin orientations of the Ising model to be replaced by a continuum of spin orientations in an n-dimensional spin space—the n-vector model). The $n = 0$ station on the East-West Metro line corresponds to the self-avoiding random walk (SAW). Two additional stations on this line have the appealing feature that they correspond to models that are exactly soluble even for three spatial dimensions ($d = 3$): $n = -2$ (random walk model) and $n = \infty$ (the spherical model). The $Q = 1$ 'station' on the *North-South* Metro line corresponds to percolation and the $Q = 3$ station to a set of adsorption problems such as krypton on graphite

The angular brackets in (5.35) indicate that the summation is over all pairs of nearest-neighbor sites $\langle ij \rangle$. The interaction energy of a pair of neighboring parallel spins is $-J$, so that if $J > 0$, the system should order ferromagnetically at $T = 0$.

The second generalization of the Ising model is the n-vector model. Each spin variable

$$S_i = (S_{i1}, S_{i2}, \ldots, S_{in}) \tag{5.37}$$

is an n-dimensional unit vector

$$\sum_{\alpha=1}^{n} S_{i\alpha}^2 = 1, \tag{5.38}$$

capable of taking on a *continuum of orientations*. Spin S_i localized on site i interacts isotropically with spin S_j localized on site j, so two neighboring spins contribute an amount $-J S_i \cdot S_j$ (with the scalar product of the two spins) to the energy. Thus the total energy of a spin configuration is

$$\varepsilon(n) = -J \sum_{\langle ij \rangle} S_i \cdot S_j . \tag{5.39}$$

The key parameter in the Potts model is Q (the number of different discrete orientations of the spin variables), just as the key parameter in the n-vector model is n (the dimension of the spin S_i). Together the Potts and n-vector models are sufficient to describe the behavior of a wide variety of systems near their critical points, and as a result immense attention has been focused on understanding these models.

For dimensions above a critical dimension d_c, the classical 'mean field' theory of Sect. 4.3.9 provides an adequate description of critical-point exponents and scaling functions, whereas for $d < d_c$, the classical theory breaks down in the immediate vicinity of the critical point because statistical fluctuations neglected in the classical theory become important. The case $d = d_c$ must be treated with great care; usually, the classical theory 'almost' holds, and the modifications take the form of weakly singular corrections.

For the n-vector model $d_c = 4$. Different values of d_c are usually found for multicritical points, such as occur when lines of critical points intersect. For example, $d_c = 3$ for a point where three critical lines intersect, and $d_c = 8/3$ for a fourth-order critical point. For a uniaxial ferromagnet or ferroelectric formed of interacting classical dipoles, $d_c = 3$; $LiTbF_4$ is one realization. In fact, $d_c = 2$ for certain structural phase transitions, such as that occurring in $PrAlO_3$.

In the models we have been considering, linear polymers can be thought of as linear clusters on a lattice. Similarly, branched polymers can be thought of as branched clusters. Such clusters are often called *lattice animals*, because they represent all the possible shapes that can be formed out of the constituent elements. Thus linear lattice animals that do not self-intersect (i.e., are loopless) are just the SAWs we discussed above. However, in general, lattice animals may branch and may form loops. Equation (5.34) may also be applied to lattice animals. The fractal dimension

Table 5.1 Comparison of some of the scaling properties of (a) self-avoiding walks (which model linear polymers), (b) lattice animals (which model branched polymers), and (c) percolation (which models gelation). The first line gives d_c, the critical dimension. The second line gives d_f, the fractal dimension, for $d \geq d_c$. The third line gives d_f^{RG}, the prediction of renormalization group expansions, for $d \leq d_c$ to first order in the parameter $\varepsilon = d_c - d$. The fourth and fifth lines give the results for dimensions three and two respectively

	(a) SAW	(b) Lattice animal	(c) Percolation
d_c	4	8	6
$d_f (d \geq d_c)$	2	4	4
$d_f^{RG} (d \leq d_c)$	$2(1 - \frac{1}{8}\varepsilon)$	$4(1 - \frac{1}{9}\varepsilon)$	$4(1 - \frac{5}{42}\varepsilon)$
$d_f (d = 3)$	≈ 1.7	2 (exact)	≈ 2.5
$d_f (d = 2)$	4/3 (exact)	≈ 1.6	91/48 (exact)

of a *random* branched object (without any restrictions) is $d_f = 4$. Hence we expect $d_c = 8$ for branched polymers, using an argument analogous to the argument for linear polymers that leads from (5.34) to the result $d_c = 4$ (Table 5.1).

A remarkable fact is that certain limiting cases of the Potts and n-vector models have a direct relation to geometrical objects that are fractal, and so these limits provide an intriguing connection between 'Physics & Geometry'. *Percolation*, e.g., is a simple geometrical model in which we study clusters formed when a fraction p of the bonds of a lattice are occupied randomly. As shown in Fig. 5.7, above a threshold value p_c ($= 1/2$ in this lattice) a subset of these bonds form a macroscopic connected object called the *infinite cluster*, and the properties of the percolation model in the vicinity of p_c are not unlike the properties of a system of cross-linking polymers in the vicinity of the gelation transition. The statistical properties of percolation can be recovered from the Q-state Potts model if we carefully form the limit $Q \to 1$. In this correspondence, the variable $p - p_c$ in percolation corresponds to the variable $T - T_c$ in the magnetic system.

Similarly, if we carefully form the $n \to 0$ limit of the n-vector model, then we recover the statistical properties of the SAW. In this correspondence, it turns out that the inverse mass M^{-1} in the polymer system corresponds to $T - T_c$ in the magnetic system. Thus the limit of large molecular weight corresponds to a critical point; we say that a growing polymer chain exhibits the phenomenon of *self-organized criticality* because as it grows it approaches a critical point. As a result, we expect quantities such as the polymer diameter to be characterized by universal behavior. In addition, the existence of a phase transition allows us to apply to the SAW problem modern techniques as the renormalization group. The use of fractal geometry (which concerns the limit $M \to \infty$) becomes relevant to studying materials near their critical points (which concern the asymptotic limit $T \to T_c$).

Theoretical physicists—for all their well-honed mathematical skills—are totally incapable of solving simply-defined models such as the Ising model or the SAW problem for the case of a three-dimensional ($d = 3$) system. However they can invent bizarre spin dimensionalities which do yield to exact solution in $d = 3$ and so provide

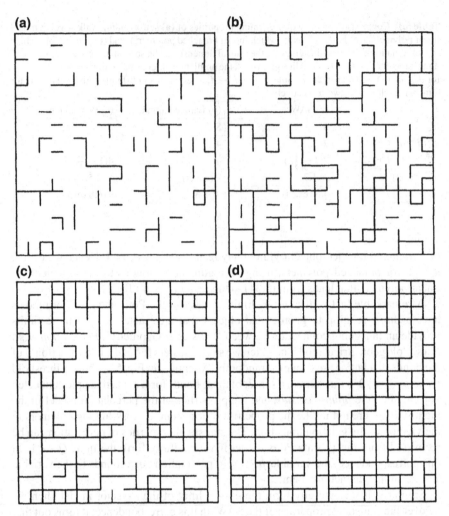

(a) **(b)** **(c)** **(d)**

Fig. 5.7 The phenomenon of bond percolation: a finite section of an infinite 'fence', in which a fraction p of the links is conducting while the remaining fraction $1 - p$ is insulating. Four choices of the parameter p are shown, **a**, $p = 0.2$; **b**, $p = 0.4$; **c**, $p = 0.6$; and **d**, $p = 0.8$

useful 'anchor points' with which to compare the results of various approximation procedures. For example, for $n = -2$ the n-vector model is found to provide the same statistical properties as for the simple unbiased random walk (the limiting case of a *non-interacting* polymer chain). In the limit $n \to \infty$ we recover a model—termed the spherical model—which has the important features of being exactly soluble for all spatial dimensions d, as well as being useful in describing the statistical properties of the Bose-Einstein condensation.

5.9 Fractal Aggregates

We began our dessert by forming a simple non-random fractal aggregate, the Sierpinski gasket. We shall end the dessert by describing one of the most popular current models for random fractal aggregates, diffusion limited aggregation (DLA).

Like many models in statistical mechanics, the rule defining DLA is simple. At time 1, we place in the center of a computer screen a black pixel, and release a random walker from a large circle surrounding the black pixel. The four perimeter sites have an equal a priori probability p_i to be stepped on by the random walker (Fig. 5.8a), so we write

$$p_i = \frac{1}{4} \quad (i = 1, \ldots, 4). \tag{5.40}$$

The rule is that the random walker remains fixed at a perimeter site if and when it ever lands on the perimeter site—thereby forming a cluster of mass $M = 2$. There are $N_p = 6$ possible sites, henceforth called *growth sites* (Fig. 5.8b), but now the probabilities are *not* all identical: each of the growth sites of the two tips has growth probability $p_{max} \cong 0.22$, while each of the four growth sites on the sides has growth probability $p_{min} \cong 0.14$. Since a site on the tip is 50% more likely to grow than a site on the sides, the next site is more likely to be added to the tip—it is like capitalism in that 'the rich get richer'. One of the main features of recent approaches to DLA is that instead of focusing on the tips who are 'getting richer', we can focus on the fjords who are 'getting poorer'—which is a realization in Nature of the familiar experience that 'Once you get behind you stay behind!' The topological scale-free

(a) t=1   (b) t=2

Fig. 5.8 a Square lattice DLA at time $t = 1$, showing the four growth sites, each with growth probability $p_i = 1/4$. **b** DLA at time $t = 2$, with 6 growth sites, and their corresponding growth probabilities p_i

Barabási-Albert networks show the same principle that the rich get richer[4] while economic reality is more complicated.[5]

Just because the third particle is *more likely* to stick at the tip does not mean that the next particle *will* stick on the tip. Indeed, the most that we can say about the cluster is to specify the *growth site probability distribution*—i.e., the set of numbers,

$$\{p_i\} \quad i = 1, \ldots, N_p, \tag{5.41}$$

where p_i is the probability that perimeter site ("growth site") i is the next to grow, and N_p is the total number of perimeter sites ($N_p = 4, 6$ for the cases $M = 1, 2$ shown in Fig. 5.5a, b respectively). The recognition that the set of $\{p_i\}$ gives us essentially the *maximum* amount of information we can have about the system is connected to the fact that tremendous attention has been paid to these p_i—and to the analogs of the p_i in various closely-related systems.

If the DLA growth rule is simply iterated, then we obtain a large cluster characterized by a range of growth probabilities that spans several orders of magnitude—from the tips to the fjords. The cover shows such a large cluster, where each pixel is colored according to the time it was added to the aggregate. From the fact that the 'last to arrive' particles (green pixels) are never found to be adjacent to the 'first to arrive' particles (white pixels), we conclude that the p_i for the growth sites on the tips must be vastly larger than the p_i for the growth sites in the fjords.

Until relatively recently, most of the theoretical attention paid to DLA has focused on its fractal dimension. Although we now have estimates of d_f that are accurate to roughly 1%, we lack any way to *interpret* this estimate. This is in contrast to both the $d = 2$ Ising model and $d = 2$ percolation, where we can calculate the various exponents *and* interpret them in terms of 'scaling powers'. What we can interpret, however, is the distribution function $D(p_i)$ which describes the histogram of the number of perimeter sites with growth probability p_i. The key idea is to focus on how this distribution function $D(p_i)$ *changes* as the cluster mass M increases. The reason why this approach is fruitful is that the $\{p_i\}$ contain the maximum information we can possibly extract about the dynamics of the growth of DLA. Indeed, specifying the $\{p_i\}$ is analogous to specifying the four 'growth' probabilities $p_i = 1/4$ [$i = 1, \ldots, 4$] for a random walker on a square lattice.

The set of numbers $\{p_i\}$ may be used to construct a histogram $\mathcal{D}(\ln p_i)$. This distribution function can be described by its *moments*,

$$Z_\beta = \sum_{\ln p} \mathcal{D}(\ln p) e^{-\beta(-\ln p)}, \tag{5.42}$$

which is a more complex way of writing

---

[4] A.L. Barabási, *Linked: The New Science of Networks*, Perseus Books Group, Cambridge MA, 2002.

[5] T. Piketty, *Capital in the Twenty-First Century*. Belknap-Harvard University Press, Cambridge MA, 2014.

$$Z_\beta = \sum_i p_i^\beta .$$ (5.43)

It is also customary to define a dimensionless 'free energy' $F(\beta)$ by the relation

$$F(\beta) = -\frac{\log Z_\beta}{\log L} .$$ (5.44)

which can be written in the suggestive form

$$Z_\beta = L^{-F(\beta)} .$$ (5.45)

The form (5.42) as well as the notation used suggests that we think of β as an *inverse temperature*, $-\ln p / \ln L$ as an *energy*, and Z_β as a *partition function*. The notation we have used is suggestive of thermodynamics. Indeed, the function $F(\beta)$ has many of the properties of a free energy function—for example, it is a convex function of its argument and can even display a singularity or 'phase transition'. However for most critical phenomena problems, exponents describing moments of distribution functions are linear in their arguments, while for DLA $F(\beta)$ is not linear—we call such behavior *multifractal*. Multifractal behavior is characteristic of random multiplicative processes, such as arise when we multiply together a string of random numbers, and can be interpreted as partitioning a DLA cluster into fractal subsets, each with its own fractal dimension (Plate 1).

Our dessert is now finished, so let us take a little exercise to work off the calories. Our exercise takes the form of a simple *hands-on* demonstration that enables us to actually 'see with our eyes' (a) that DLA is a fractal and (b) that its fractal dimension is approximately 1.7. We begin with a large DLA cluster (Plate 1), and cut out from three sheets of scrap paper holes of sizes $L = 1, 10, 100$ (in units of the pixel size). Now cover the fractal with each sheet of paper, and estimate the fraction of the box that is occupied by the DLA. This fraction should scale in the same way as the density $\rho(L) = M(L)/L^2$ which, from (5.4), decreases with increasing length scale as $\rho(L) = AL^{d_f-2}$. Now (5.4) is mathematically equivalent to the functional equation

$$\rho(\lambda L) = \lambda^{d_f - d} \rho(L) .$$ (5.46)

For our exercise, $\lambda = 10$ and we find

$$\rho(L) \approx \begin{cases} 1 & L = 1 \\ 1/2 & L = 10 \\ 1/4 & L = 100 . \end{cases}$$ (5.47)

Here the result of (5.47),

$$\rho(10L) \approx \frac{1}{2}\rho(L) ,$$ (5.48)

convinces us that $10^{d_f-2} \approx \frac{1}{2}$, leading to $d_f - 2 \approx \log_{10} \frac{1}{2} = -0.301$, or

$$d_f \approx 1.70 \,. \tag{5.49}$$

This crude estimate agrees with a value from computer simulations,

$$d_f = 1.712 \pm 0.003 \,, \tag{5.50}$$

based on clusters with 5×10^7 particles (P. Ossadnik, Physica A **195**, 319 (1993)).

5.10   Fractals in Nature

The reader has savored the meal, indulged himself on the dessert, and is now entitled to a little fantasy before falling asleep for the night. Accordingly, we shall describe in this final section some of the situations in Nature where fractal phenomena arise and wax philosophical about exactly how much theoretical physics might hope to contribute to our understanding of these phenomena.

Everyone has seen many fractal objects—probably at an early stage in life. Perhaps we once photographed scenery from the back of a train and noticed that the photograph looked the same at all stages of enlargement (Fig. 5.9). Perhaps we noticed

Fig. 5.9 Schematic illustrations of scale invariance for a blow-up of the central portion of a photograph from the rear of a train in a flat terrain like Oklahoma

Fig. 5.10 Typical retinal neuron and its fractal analysis. The correlation function $C(r)$ in scales in the same fashion as the density, given by (5.4). (*See* F. Caserta, H.E. Stanley, W. Eldred, G. Daccord, R. Hausman, and J. Nittmaim, "Physical Mechanisms Underlying Neurite Outgrowth: A Quantitative Analysis of Neuronal Shape", Phys. Rev. Lett. **64**, 95 (1990))

that the Metro of Paris has a fractal structure in the suburbs (M. Benguigui and M. Daoud, Geographical Analysis **23**, 362 (1991). Perhaps we saw that snow crystals all have the same pattern, each part of a branch being similar to itself. In fact, to 'see' something at all—fractal or non-fractal—requires that the nerve cells in the eye's retina must send a signal, and these retinal nerve cells are themselves fractal objects (Fig. 5.10).

There are many *caveats* that we must pay heed to. To be fractal implies that a part of the object resembles the whole object, just as the branches of a DLA look similar to the whole structure and also similar to the sub-branches. The Sierpinski gasket shows this self-similarity exactly, whereas for DLA and other random fractals this self-similarity is only statistical. Fractal objects in Nature are random fractals, so the self-similarity we discover by enlarging the middle section of Fig. 5.9 is replaced by a self-similarity obtained only by averaging together many realizations of the same object.

Fig. 5.11 A Sierpinski
gasket that is self-similar
(fractal) on small length
scales, but becomes
homogeneous (non-fractal)
on large length scales

The second *caveat* is that fractals in Nature are not fractal on all length scales. There is a range of length scales, followed by an inevitable crossover to homogeneous behavior. We can indicate this fact using the Sierpinski gasket model of Sect. 5.1 by simply starting, after n stages, to aggregate exact copies of the object, so that asymptotically one obtains a homogeneous object made up of units each identical to the n-stage Sierpinski gasket (Fig. 5.11). The result is a crossover phenomenon. This example is instructive, because the resulting behavior is analogous to what is usually found in Nature: real objects do not remain fractal for all scales, but instead are fractal over typically a factor of 10 or 100 in length scale. The fact that real objects in Nature do not remain fractal on all length scales does not make them any less interesting—there can even be useful information in the value of the length scale on which the crossover to homogeneous behavior occurs.

With these caveats, however, it is a fact that fractals abound in Nature. In fact, almost any object for which randomness is the basic factor determining the structure will turn out to be fractal over some range of length scales—for much the same reason that the simple random walk is fractal: there is nothing in the microscopic rules that can set a length scale so the resulting macroscopic form is 'scale-free'...

scale-free objects obey power laws and lead to functional equations of the form of (5.19) and (5.29).

Today, there are roughly of order 10^3 fractal systems in Nature, though a decade ago when Mandelbrot's classic was written, many of these systems were not known to be fractal. These include examples of relevance to a wide variety of fields, ranging from geological chemistry (Plate 3) and fracture mechanisms (Plate 4) on the one hand, to fluid turbulence (Plate 5) and the "molecule of life"—water (Plate 6)—on the other. DLA alone has about 50 realizations in physical systems. DLA models aggregation phenomena described by a Laplace equation ($\nabla^2 \Pi(r, t) = 0$) for the probability $\Pi(r, t)$ that a walker is at position r and time t. More surprising is the fact that DLA describes a vast range of phenomena that at first sight seem to have nothing to do with random walkers. These include fluid-fluid displacement phenomena ("viscous fingers"), for which the pressure P at every point satisfies a Laplace equation (Plates 7–10). Similarly, dielectric breakdown phenomena, chemical dissolution (Plate 11), electrodeposition, and a host of other phenomena may be members of a suitably-defined *DLA universality class*. If anisotropy is added, then DLA describes dendritic crystal growth and snowflake growth (Fig. 5.12). The dynamics of DLA growth can be studied by the multifractal analysis discussed above, or by decomposing a DLA cluster into active tips connected to the central seed by a "skeleton" from which emanate a fractal hierarchy of branches whose dynamics resembles $1/f$ noise (Plate 12).

Recently, several phenomena of *biological* interest have attracted the attention of DLA *aficionados*. These include the growth of bacterial colonies, the retinal vasculature, and neuronal outgrowth (Fig. 5.10). The last example is particularly intriguing: if evolution indeed chose DLA as the morphology for the nerve cell, then can we understand 'why' this choice was made? What evolutionary advantage does a DLA morphology convey? Is it significant that the Paris Metro evolved with a similar morphology or is this fact just a coincidence? Can we use the answer to these questions to better design the next generation of computers? These are important issues that we cannot hope to resolve quickly, but already we appreciate that a fractal object is the most efficient way to obtain a great deal of intercell 'connectivity' with a minimum of 'cell volume', so the key question is 'which' fractal did evolution select, and why?

It is awe-inspiring that remarkably complex objects in Nature can be quantitatively characterized by a single number, d_f. It is equally awe-inspiring that such complex objects can be described by various models with extremely simple rules. It is also an intriguing fact that even though no two natural fractal objects that we are likely to ever see are identical, nonetheless every DLA has a generic 'form' that even a child can recognize. The analogous statement holds for many random structures in Nature. For example no two snowflakes are the same yet every snowflake has a generic form that a child can recognize.

Perhaps most remarkable to a student of theoretical physics is the fact that simple geometrical models—with no Boltzmann factors—suffice to capture features of real statistical mechanical systems such as those discussed in Chap. 4. What does this mean? If we understand the essential physics of an extremely robust model, such as the Ising model, then we say that we understand the essential physics of the

Fig. 5.12 a A typical snow crystal, **b** a DLA simulation, **c** comparison between the fractal dimensions of **a** and **b** obtained by plotting the number of pixels inside an $L \times L$ box logarithmically against L. The same slope, $d_f = 1.85 \pm 0.06$, is found for both. The experimental data extend to larger values of L, since the digitizer used to analyze the experimental photograph has 20,000 pixels while the cluster has only 4000 sites. (*See* J. Nittmann and H.E. Stanley, "Non-Deterministic Approach to Anisotropic Growth Patterns with Continuously Tunable Morphology: The Fractal Properties of Some Real Snowflakes", J. Phys. A **20**, L1185 (1987))

complex materials that fall into the universality class described by the Ising model. In fact, by understanding the pure Ising model, we can even understand most of the features of *variants* of the Ising model (such as the n-vector model) that may be appropriate for describing even more complex materials. Similarly, we feel that if we can understand DLA, then we are well on our way to understanding *variants* of DLA, such as DLA with noise reduction (Plate 13), the screened growth model (Plate 14), ballistic deposition (Plate 15), and cluster-cluster aggregation (Plate 16).

Chapter 6
Dynamical Systems and Chaos

Abstract Dynamical systems theory offers a different, geometrical view for describing the evolution of a system. It leads to the puzzling notions of deterministic chaos and universal routes to chaos.

Dynamical systems theory proposes a new point of view to describe the evolution of a system S. It appeared after the Newtonian (Sect. 1.1) and Lagrangian (Sect. 1.3) viewpoints, since it is rooted in the work of Poincaré (1892), Birkhoff (1912, 1931) and Lyapunov (1916). The most remarkable—and remarked—advances, about forty years ago, have been the development of the fascinating notion of *deterministic chaos* and the unravelling of *universal routes* to chaos. This theory may seem more formal and distant from reality than the familiar mechanics presented in Chap. 1. However, after defining the basic notions, we will see how it applies to real evolutions. We will enlighten specific issues that cannot be solved in an other framework.

6.1 Basic Notions and Framework

6.1.1 Phase Space

The first step to formally represent the evolution of a real system[1] S is to *quantitatively* describe the state of S, by means of some quantities whose values are real numbers. For instance, the qualitative notion of color has to be replaced by that of wavelength. Moreover, it is not possible to describe all the features of S. Some of them are besides of no influence on the dynamics. Archimedes had surely noticed that the color of a body does not affect its buoyancy. The physicist thus begins by *choosing* a set of variables whose value at time t, denoted in a compact way $x(t)$ whatever the number of variables and their meaning, describes the state of S at this moment. This choice

---

[1] The *real system* S and the *dynamical system* modeling the evolution of a few quantities characterizing the state of S should not be confused.

© Springer-Verlag GmbH Germany 2017

D. Stauffer et al., *From Newton to Mandelbrot*, Graduate Texts in Physics,
DOI 10.1007/978-3-662-53685-8_6

depends on the *scale* at which the physicist wants to describe S and on its intuition regarding the mechanisms that actually control the dynamics. A possible way is to start with a (too) precise model which is pruned into a minimal description. The opposite way, starting from a (too) simple model which is later refined, is often more efficient and more robust. This choice amounts to prescribe the space, termed the *phase space*, in which the point x varies. This space is nothing but the space of the states of S in the chosen representation. In this chapter, we will limit ourselves to simple cases where the components x_j of $x \in \mathcal{X}$ are real numbers or angles.

Let us consider as an example a system S composed of $N \gg 1$ particles. In a microscopic description (Sects. 1.3.3, 4.3), the phase space is the set $\mathcal{X}_{\text{micro}} = \{(q_i, p_i)_{i=1\ldots N}\} \subset \mathbf{R}^{6N}$, in which a point represents the generalized coordinates of the N particles. In a macroscopic description (Sect. 4.2), the phase space is the set $\mathcal{X}_{\text{macro}} = \{(N, P, V, T)\} \subset \mathbf{R}^4$ in which a point represents the thermodynamic observables.

Quite often, the phase space $\mathcal{X}$ comprises variables that are neither positions nor velocities but for instance concentrations or temperatures. We thus aim at describing more generally *dynamics*, not only motions.

6.1.2 Continuous-Time Dynamical Systems

A *dynamical system* in $\mathcal{X}$ is a model of evolution of the form:

$$\dot{x}(t) = V[x(t)], \tag{6.1}$$

where $\dot{}$ denotes the derivative with respect to time $t \in \mathbf{R}$. The unknown function is $x(t)$, with values in $\mathcal{X}$. V is a given function in[2] $\mathcal{X}$. Such an evolution is an evolution *with no memory*, meaning that the knowledge of the state $x(t_0)$ at one time t_0 determines all the subsequent states of S—at least in principle since it is not always possible to solve the differential equation (6.1). We assume that V is smooth enough.[3] The solution of (6.1) with value x_0 at time $t = 0$ is called the *trajectory* of x_0. Note that it is a trajectory *in the phase space*. The dynamical system (6.1) is said to be *autonomous* when V does not depend explicitly on time. This means that the evolution law is stationary i.e. invariant upon a temporal translation. It is always possible to come down to such a situation by extending the phase space hence we will consider only the case of autonomous systems.

---

[2]Rigorously, V is defined on $\mathcal{X}$ and its value $V(x)$ belongs not to $\mathcal{X}$ but to the tangent space to $\mathcal{X}$ in x; the distinction is important for instance when $\mathcal{X}$ is a circle and x an angular variable. In mathematical terms, such a function V is called a *vector field* on $\mathcal{X}$.

[3]It is enough that V is C^1 in $\mathcal{X}$, that is, differentiable with a continuous differential $x \mapsto DV(x)$, to ensure the existence and unicity of the solutions. The rationale behind such a regularity assumption is that we are interested in physical models intending to account for realistic dynamics. Singular evolution equations may have several solutions for a given initial conditions, or even worst, no solutions. We here exclude such unphysical behaviors.

6.1.3   Flows and Phase Portraits

Solutions of (6.1) can be represented using a single function f defined on $\mathbf{R} \times \mathcal{X}$ and taking its values in $\mathcal{X}$. It is termed the *flow* of the dynamical system. The value $f(t, x_0)$, also denoted $f_t(x_0)$, is the point of $\mathcal{X}$ describing the state of $\mathcal{S}$ at time t when it starts in state x_0 at time 0. It satisfies:

$$f_{t'} \circ f_t = f_{t'+t} \quad \text{(for any } t, t' \in \mathbf{R}\text{).} \tag{6.2}$$

This relation expresses the invariance of (6.1) upon a change of time origin when V does not depend explicitly on time: $\mathcal{S}$ reaches the same point at time $t' + t$ if it starts in state x_0 at time 0 or in state $f_t(x_0)$ at time t. The evolution law (6.1) is *reversible* insofar as the flow is defined for any $t \in \mathbf{R}$ and $f_{-t} = (f_t)^{-1}$ (by taking $t' = -t$ in (6.2)). The flow offers two viewpoints on the evolution of $\mathcal{S}$:

- either like a bundle of trajectories $[t \mapsto f_t(x_0)]$ parameterized by their initial condition $x_0 \in \mathcal{X}$ (like a bunch of spaghetti),
- or as a sequence of instantaneous snapshots $[x_0 \mapsto f_t(x_0)]$, describing the deformation along time of the image formed by the points of $\mathcal{X}$ (like a spliced fruit cake).

Equation (6.2) ensures that $\mathcal{S}$ is in state $f_t(x_0)$ at time $t + t_0$ if it starts in state x_0 at time t_0, *whatever t_0 is*. This property implies that two trajectories cannot cross unless they are identical. A *unique* trajectory, defined from $t = -\infty$ to $t = +\infty$, thus passes through a given point of $\mathcal{X}$ at $t = 0$. The covering of $\mathcal{X}$ by these non-intersecting curves is termed the *phase portrait* of the dynamical system. A phase portrait does not give any information on the motion along the trajectories since it displays only their path without keeping track of the time course along a path. In return, it can be obtained without having to solve the evolution equations. In two dimensions (so that the drawing is tractable), $x = (x_1, x_2)$ and $V = (V_1, V_2)$. The tangent in x to the trajectory passing through this point has a slope:

$$\alpha(x) = \frac{dx_2}{dx_1} = \frac{\dot{x}_2}{\dot{x}_1} = \frac{V_2(x)}{V_1(x)}. \tag{6.3}$$

Without any integration, it is thus possible to visualize the flow, point-wise, exactly like field lines are visualized, point-wise, using iron filings. We will present below in more detail how to draw a phase portrait (Sect. 6.2.4).

6.1.4   Insights from Dynamical Systems Theory

At this point an objection can be raised: Equation (6.1) is nothing but an ordinary differential equation of order 1, an entity studied for long by mathematicians without waiting for advice from physicists! The specificity of dynamical systems theory does not lie in the investigated object, but rather in the (physically motivated) questions addressed about it and the methodology developed for providing answers.

The main issue faced to a dynamical system is to predict the asymptotic behavior $t \to \infty$ of the trajectories. This limit $t \to \infty$ corresponds in practice to long-term observations. The name *asymptotic dynamics* refers to the stationary regime (also called steady state) that is observed once transients has relaxed. More precise questions are for instance:

- Which initial conditions lead to identical long-term behaviors? and which ones will on the contrary yield dramatically different behaviors?
- Can we predict the phase space region in which the trajectory is confined, if any?
- Can we estimate the rate of convergence to an equilibrium state?
- How is modified the phase portrait when V is slightly perturbed?

These are crucial questions in face of which we might feel helpless when we do not known to solve (6.1) and give explicit solutions for any $t \in [0, \infty]$. Numerical methods do not solve the difficulty, because they provide solutions (moreover approximate ones) only over a finite duration: numerical integration cannot give any insight on the limiting behavior for $t \to \infty$, unless preliminary theoretical results provide a guide rope. In this latter case, numerical methods can be developed to compute asymptotic quantities whose existence and meaning are known in advance (see below the example of Lyapunov exponents). In order to answer the above questions, novel methods and even novel theoretical notions are necessary. Actually a *change of viewpoint* is required: to better focus on the asymptotic dynamics, one chooses to ignore details about the transients, e.g. the explicit expression of the trajectories on finite time intervals $[0, T]$. This is the core and the specificity of the theory of dynamical systems.

6.1.5 Some Examples

Example 1: Mechanical System
Newtonian mechanics presented in Sect. 1.1 can be reformulated within the framework of dynamical systems theory. The equation describing the evolution of a (classical) mechanical system S with mass m and a single degree of freedom x is written $m\ddot{x} = F(x, \dot{x})$, where F is the resultant of the forces experienced by S. Introducing the additional variable $y = m\dot{x}$ turns this equation of evolution into a dynamical system in the plane:

$$\begin{cases} \dot{x} = y/m \\ \dot{y} = F(x, y/m). \end{cases} \tag{6.4}$$

The extension to the case where S has d mechanical degrees of freedom is straightforward: the corresponding dynamical system is then of (necessarily even) dimension $2d$. The trick can be generalized to any ordinary differential equation $d^n x/dt^n = F(x, dx/dt, \ldots, d^{n-1}x/dt^{n-1})$ of order n in $\mathbf{R}^d$ by introducing $n - 1$

auxiliary variables $y_j = d^j x / dt^j$ ($j = 1, \ldots, n-1$), which yields a dynamical system in $\mathbf{R}^{nd}$:

$$\begin{cases} \dot{y}_j = y_{j+1} & j = 0, \ldots, n-2 \\ \dot{y}_{n-1} = F(y_0, y_1, \ldots, y_{n-1}). \end{cases} \quad (6.5)$$

Example 2: Hamiltonian Systems
The mechanical system S is termed *Hamiltonian* when there exists generalized coordinates $q = (q_1 \ldots q_d)$ an $p = (p_1 \ldots p_d)$, and a function $H(q, p)$ such that its evolution is described by the canonical equations presented in Sect. 1.3:

$$\begin{cases} \dot{q}_j = \frac{\partial H}{\partial p_j}(q, p) \\ \dot{p}_j = -\frac{\partial H}{\partial q_j}(q, p) \end{cases} \quad j = 1, \ldots, d. \quad (6.6)$$

When the Hamiltonian H does not depend explicitly on time ($\partial_t H = 0$), the total energy of S is a constant quantity E: $\forall t$, $H(q_t, p_t) = E$. The system is then said to be *conservative* and Liouville theorem[4] ensures that the volume element $d^d q d^d p$ is conserved along time: the evolution changes the shape of the phase space regions without changing their volume.

If S possesses d invariant quantities $J_1 \ldots J_d$ (as many as the number of mechanical degrees of freedom), the system is said to be *integrable*. It is then possible to show that there exists a coordinate change $(q, p) \mapsto (J, \theta)$ such that S satisfies Hamilton equations with respect to the new variables and the same Hamiltonian $\tilde{H}(J, \theta) = H(q, p)$. Such a coordinate change preserving Hamilton equations is termed a *canonical transformation*. Variables $J = (J_1 \ldots J_d) \in \mathbf{R}^d$ are called the *actions* (or action variables) and $\theta = (\theta_1 \ldots \theta_d) \in \mathbf{T}^d$ the *angles* (or angular variables). Functions q, p and $\tilde{H}$ are 2π-periodic with respect to each angular variable θ_j (whence their name).

Given the d canonical equations $\dot{J}_j = -\partial \tilde{H} / \partial \theta_j$, the fact that J is constant along time ($\dot{J} = 0$) implies that $\tilde{H}$ depends only on J. The equations of motion thus reduce to the trivial dynamical system:

$$\dot{\theta}_j = \frac{\partial \tilde{H}}{\partial J_j}(J) \equiv \omega_j(J) = \text{const.} \quad j = 1, \ldots, d. \quad (6.7)$$

Integration is straightforward. This simplification is the rationale motivating the change of variables $(q, p) \mapsto (J, \theta)$. Inverting this transformation to recover position variables yields a *quasi-periodic* motion

$$q_j(t) = Q_j(\omega_1 t + \theta_1^0, \ldots, \omega_d t + \theta_d^0) \quad j = 1, \ldots, d, \quad (6.8)$$

---

[4]In order to better understand this theorem, the reader can compute the determinant of the Jacobian matrix of the transformation $(q, p) \mapsto (q + \dot{q} \Delta t, p + \dot{p} \Delta t)$ which approximately describes (at the lowest order in Δt) the evolution over a time step Δt. The rigorous proof for the exact evolution law (in continuous time) yields the same conclusion.

where the functions $(Q_j)_{j=1...d}$ are 2π-periodical with respect to each of their arguments. Angles $\theta_1^0, \ldots, \theta_d^0$ are integration constants. The quasi-periodic motion is periodic if the ratios of its frequencies are all rational: $\omega_j/\omega_1 = r_j/r_1$, where $(r_j)_j$ are integers. The period is then $T = 2\pi r_1/\omega_1$. There is more to say when the system is not integrable. We will come back to this topic in Sect. 6.6.4.

Example 3: Chemical Systems

Equations describing at our scale the kinetics of chemical reactions are dynamical systems, with variables the concentrations of reactants and products. For instance, the reaction:

$$A + 2B \xrightarrow{k} 2C + D \tag{6.9}$$

is described by the following dynamical system (with obvious notations):

$$\begin{cases} \dot{a} = -kab^2 \\ \dot{b} = -2kab^2 \\ \dot{c} = 2kab^2 \\ \dot{d} = kab^2. \end{cases} \tag{6.10}$$

We see here an example of reduction of the "a priori" phase space $\mathcal{X} = \{(a, b, c, d)\}$. Indeed, $-\dot{b} = \dot{c} = 2\dot{d} = -2\dot{a}$; in particular $a + b + c + d \equiv 1$. Among the four variables seemingly required to describe the system, there is only one independent variable, say a, whose evolution is ruled by

$$\dot{a} = -ka(b_0 - 2a_0 + 2a)^2 \tag{6.11}$$

(a_0 and b_0 are constant and fixed by the initial conditions). Similarly, the so-called *self-catalytic* reaction (when the presence of species A enhances its own production): $A + B \xrightarrow{k} 2A$ in a closed chemical reactor is described by the one-dimensional dynamical system: $\dot{a} = -ka(1 - a)^2$ since then $a + b = $ const. normalized to 1. More complex reactions, fulfilling less conservation laws are described by dynamical systems with multidimensional phase spaces. This is for instance the case in open reactors, where reactants are continuously injected and products extracted. All these dynamical systems satisfy the following property (required in order to interpret the variables in terms of concentrations): *if the variables $(a, b, \ldots)$ are positive at $t = 0$, they remain positive at all times.*

The reader has now acquired enough knowledge to prove this result, at least in the above examples.[5]

Example 4: Lorenz Equations

This historical example has been introduced by Lorenz in 1963 to describe atmospheric dynamics. The first step leading to this model is to truncate and

---

[5]$a \equiv 0$ is a solution. As any trajectory, it behaves as an impenetrable barrier trapping the other trajectories either in the region $a > 0$ or in the region $a < 0$ according to the initial condition. This reasoning extends to the other variables.

simplify hydrodynamic equations (spatiotemporal partial differential equations, see Sect. 1.5.4) ruling air motion in the atmosphere. The second step is to look for solutions with a specified spatial dependence (guessed from the qualitative understanding of the equations and the expertise of Lorenz) parameterized by three unknown real-valued functions $X(t)$, $Y(t)$, $Z(t)$ evolving in time. This yields a dynamical system in $\mathbf{R}^3$:

$$\begin{cases} dX(t)/dt = \sigma(Y - X) \\ dY(t)/dt = rX - Y - XZ \\ dZ(t)/dt = XY - bZ, \end{cases} \tag{6.12}$$

where σ, r et b are constant. At the price of some approximations, this derivation reduces an infinite-dimensional system (the initial phase space is composed of functions of the space variables) to a system in dimension 3. This Lorenz system played an important role in the discovery of *chaos* and we will come back to it in Sect. 6.5.

6.2 Fixed Points and Linear Stability Analysis

6.2.1 Fixed Points and Stability Matrix

A *fixed point* of the dynamics is a point $x^* \in \mathcal{X}$ such that $V(x^*) = 0$. It is obvious that the constant function of time $x(t) \equiv x^*$ is a solution: it is a trajectory reduced to a single point. This corresponds to an *equilibrium state* since if S is in state x^* at the initial time, it remains in this state indefinitely. The natural question for a physicist is to investigate the *stability properties* of this equilibrium state. The equilibrium is *stable* if S tends to return to it after a small disturbance, or on the contrary *unstable* if the initial discrepancy tends to increase. This is a familiar notion. Let us consider a pendulum, described by the angle θ between the pendulum and the downwards vertical direction. The position $\theta = 0$ (ball down) is stable whereas the symmetric position $\theta = \pi$ (ball up) is unstable. When the dimension of $\mathcal{X}$ is higher than 1, stability is a richer notion as the equilibrium can be at the same time stable or unstable, according to the direction $x_0 - x^*$ in which S is moved away from x^*.

In order to determine how S evolves from a state x_0 close to the equilibrium state, that is, to describe the behavior of the trajectories in the neighborhood of x^*, the first step in to perform a *linear stability analysis*. Let us give a simple example to discuss the validity and the limits of this method.

Example 5: $\dot{x} = ax(1 - x)$
The phase space is the real line, of dimension 1. To capture the evolution close to the equilibrium state $x^* = 0$, the idea is to consider only the dominant part $\dot{y} = ay$, whose solution $y_0 e^{at}$ departs from 0 if $a > 0$ and converges to 0 if $a < 0$. The comparison with the exact solution, left as an exercise, shows that here the linear approximation yields a correct understanding of the dynamics near 0.

The approach can be generalized: the linear approximation of an evolution (6.1) in the neighborhood of a fixed point x^* is obtained in a systematic way by replacing $V(x)$ by its first-order Taylor expansion around x^*:

$$\dot{y} = DV(x^*)(y - x^*). \qquad (6.13)$$

The zeroth-order is absent since $V(x^*) = 0$. Linearized evolution equation is a differential equation with constant coefficients, whose resolution is straightforward:

$$y(t) = x^* + e^{tDV(x^*)}(y_0 - x^*). \qquad (6.14)$$

Trajectories come close to x^* in the directions (eigenvectors) associated with the eigenvalues of $DV(x^*)$ having a strictly negative real part, called *stable directions* and spanning the *stable space*. On the contrary, trajectories move away from x^* in the directions (eigenvectors) associated with the eigenvalues of $DV(x^*)$ having a strictly positive real part, called *unstable directions* and spanning the *unstable space*. For this reason, $DV(x^*)$ is called the *stability matrix*. This reasoning ignore nonlinear terms and one may (should) wonder about its validity. The answer is given in the next paragraph. The case of vanishing eigenvalues appears to correspond to a specific phenomenon, bifurcation, and is treated in Sect. 6.3.

6.2.2 Flow Linearization Theorem

The issue raised by linear stability analysis is the validity of ignoring nonlinear termed near x^* and concluding about the asymptotic behavior of the solution (6.1) from the sole analysis of the linear flow ruled by (6.13). The answer is: *Yes if $DV(x^*)$ has no eigenvalues with vanishing real parts*. This fundamental result involves mathematical technicalities, and the following paragraph can be skipped during a first reading.

To intuitively explain the theorem associated with the above statement, let us get back to the Example 5. If $a \neq 0$, one defines $y = \psi(x) \equiv x/(1 - x)$. The function $y(t)$ obtained using this transformation of the exact solution $x(t)$ is a solution of $\dot{y} = ay$. Inverting $x(t) = \psi^{-1}[y(t)]$ yields the desired equivalence between the exact flow and the linearized flow.

What allowed us here to determine the behavior of $x(t)$ from the knowledge of $y(t)$ is their invertible and time-independent relationship ψ, provided $a \neq 0$, and $x(t)$ and $y(t)$ are close enough to 0.

Stated in a more general setting, the result is called the *(Hartman-Grossman) linearization theorem*. Keeping the notation $x(t)$ for the exact solution and $y(t)$ for the solution of the linearized evolution, this theorem states:

If $DV(x^*)$ has no vanishing or purely imaginary eigenvalues, there exists two neighborhoods, $\mathcal{U}_0$ and $\mathcal{U}_1$, of the fixed point x^* and an homeomorphism (continuous application with a continuous inverse function) ϕ between $\mathcal{U}_0$ and $\mathcal{U}_1$, such that $x(t) = \phi[y(t)]$ provided $y(t)$ is in $\mathcal{U}_0$, and $\phi(x^*) = x^*$. If x_0 is close enough to x^*, i.e. $x_0 \in \mathcal{U}_1$, then $y_0 = \phi^{-1}(x_0) \in \mathcal{U}_0$. It is thus possible to relate for a finite duration $t_{\max}(x_0)$ (depending on x_0) the exact trajectory $x(t)$ starting in x_0 and the solution $y(t)$ of the linearized evolution starting in y_0 (however, in general, $y_0 = \phi^{-1}(x_0) \neq x_0$). As long as $t < t_{\max}(x_0)$, an explicit form of the solution is available:

$$x(t) = \phi\left[x^* + e^{tDV(x^*)}[\phi^{-1}(x_0) - x^*]\right]. \tag{6.15}$$

The inverse application ϕ^{-1} straightens up the trajectories of (6.1) and makes them coincide with the trajectories of the linearized flow. This theorem is thus a fundamental result: without having to determine ϕ, it justifies the linear approximation of the dynamics, without which physics would be quite complicated. It also delineates situations where the approximation fails, when $DV(x^*)$ has eigenvalues with a vanishing real part. We will see in Sects. 6.2.6 and 6.3 that these situations play an important role and correspond to bifurcation points where a qualitative change occurs in the dynamics.

6.2.3 The Different Types of Fixed Points

We have just seen that if $DV(x^*)$ has no vanishing eigenvalues nor eigenvalues with a vanishing real part,[6] the behavior of the trajectories in the neighborhood of the fixed point x^* can be determined by studying only the linearized dynamics around x^*. This boils down to diagonalizing[7] the matrix $DV(x^*)$ which can be easily done at least numerically.

Let $(\lambda_j)_{j \geq 1}$ be the eigenvalues of $DV(x^*)$. Writing the initial state x_0 in the basis formed of associated eigenvectors $(v_j)_{j \geq 1}$:

$$x_0 = x^* + \sum_{j \geq 1} c_j(x_0)v_j, \tag{6.16}$$

the leading terms of the state at time t can be expressed:

$$x(t) = x^* + \sum_{j \geq 1} c_j(x_0)e^{\lambda_j t}v_j + \cdots \tag{6.17}$$

---

[6]The statement is a bit more complicated in infinite dimension, since then the spectrum of $DV(x^*)$ does not reduce to eigenvalues and the absence of eigenvalues with a vanishing real part does not guarantee the invertibility.

[7]If $DV(x^*)$ has an eigenvalue of multiplicity $m > 1$ with only one associated eigenvector, it is not diagonalizable and can only be reduced to a Jordan form. This restriction does not markedly change stability properties since the ensuing modification of (6.17) is the addition of terms $t^j e^{\eta t}$ ($j = 1, \ldots, m - 1$) to the exponential $e^{\eta t}$.

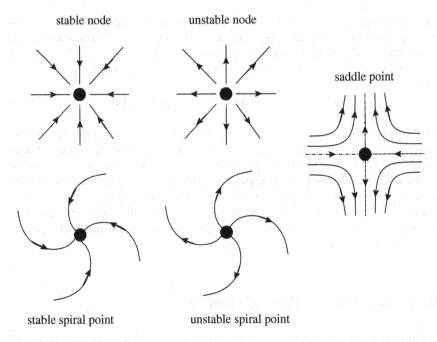

stable node   unstable node

saddle point

stable spiral point   unstable spiral point

Fig. 6.1 Types of fixed points

In this way, the stability of the fixed point can be investigated and the speed of convergence according to the direction along which the fixed point is approached can be computed. The various types of behavior in the neighborhood of a fixed point x^* can be listed according to the sign of the real part of the eigenvalues of $DV(x^*)$. We limit ourselves to the case of a phase space of dimension $d = 2$ for which graphical representation is simple (generalization to dimension $d > 2$ is straightforward). $DV(x^*)$ has then two eigenvalues λ_1 and λ_2. As illustrated in Fig. 6.1, the fixed point is:

- a *stable node* if λ_1 and λ_2 are real and strictly negative;
- an *unstable node* if λ_1 and λ_2 are real and strictly positive;
- a *stable spiral point* if λ_1 and λ_2 are complex conjugates with a strictly negative real part;
- an *unstable spiral point* if λ_1 and λ_2 are complex conjugates with a strictly positive real part;
- a *saddle* if λ_1 is strictly positive and λ_2 is strictly negative (necessarily real in this case).

In dimension $d > 2$, a fixed point is a saddle as soon as it has eigenvalues with strictly positive real part and also eigenvalues with strictly negative real part. A saddle fixed-point is neither stable nor unstable: it exhibits *jointly* an attractive influence and a repulsive influence, in different directions. The behavior of the trajectories reflects the trade-off between these two influences: trajectories experience attraction in the stable directions at short times and repulsion in the unstable directions at long times (see Example 6 below). We will come back in Sect. 6.3.3 to the delicate case where $DV(x^*)$ has eigenvalues with vanishing real part.

6.2.4 Constructing the Phase Portrait

We will now illustrate the notions introduced above and learn on an example how to extract as much information as possible about a dynamics like (6.1) without solving the equation of evolution, relying instead on elementary computation. Let us consider the following dynamical system.

Example 6
The reader is invited to take a pencil and a sheet of paper and determine the phase portrait by following the instructions of the text, before looking at the solution given at the end of the chapter in Fig. 6.10. The first, obvious feature is the specific symmetry $x \leftrightarrow y$ of the system hence of its phase portrait. As a rule, the study should always start by looking at the *symmetries* displayed by the system.

$$\begin{cases} \dot{x} = y(1-x) \\ \dot{y} = x(1-y). \end{cases} \tag{6.18}$$

By construction, $\dot{x}$ and $\dot{y}$ are state functions whose values in each point (x, y) of the phase space (here the plane) are known. Fixed points are $A = (0, 0)$ and $B = (1, 1)$. Obviously $\dot{x} = 0$ on the lines $H_0 : \{y = 0\}$ (in which A is removed) and $V_1 : \{x = 1\}$ (in which B is removed). These lines (with one point removed) are the loci of vertical tangent vectors to the trajectories, and the sign of $\dot{x}$ changes from one side to the other. In a similar way, the lines $H_1 : \{y = 1\}$ (in which B is removed) and $V_0 : \{x = 0\}$ (in which A is removed) are the loci of horizontal tangent vectors to the trajectories ($\dot{y} = 0$) and the sign of $\dot{y}$ changes from one side to the other. $\dot{x}$ and $\dot{y}$ have a constant sign within the domains delineated by these four lines. On the plot, we first draw H_0, H_1, V_0 and V_1. They delineate regions in the plane, and we represent the pair of signs associated to each region by the symbols $\nearrow$, $\searrow$, $\swarrow$ or $\nwarrow$ with obvious meaning. We also draw the oriented horizontal and vertical tangent vectors.

The phase portrait is then completed by investigating the stability properties of the fixed points A and B (that can be yet partly guess from the field of slopes) and determining the associated eigendirections. One computes:

$$DV(x, y) = \begin{pmatrix} -y & 1-x \\ 1-y & -x \end{pmatrix}, \quad \text{hence} \tag{6.19}$$

$$DV(A) = \begin{pmatrix} 0 & 1 \\ 1 & 0 \end{pmatrix} \qquad DV(B) = \begin{pmatrix} -1 & 0 \\ 0 & -1 \end{pmatrix} \tag{6.20}$$

For $t \to +\infty$, B is stable. A is a saddle point since $DV(A)$ has a stable eigenvalue $\lambda^- = -1$, associated with the eigenvector $e^- = (-1, 1)$ and and unstable eigenvalue $\lambda^+ = 1$, associated with the eigenvector $e^+ = (1, 1)$. When possible, it is interesting to delineate some trajectories, insofar as each trajectory behaves as a border that other trajectories cannot cross. In this specific example, each half of H_1 and V_1, ending in B, are trajectories. A general result for any saddle point A in dimension 2, is the existence of two particular trajectories arriving in A tangentially to the stable direction and two trajectories leaving A tangentially to the unstable direction. In our example, each trajectory is located in a different quadrant around A, leading to denote S_j that located in the j-th quadrant. By symmetry, S_1 and S_3 are half-lines of slope 1. The four trajectories $(S_j)_{j=1...4}$ are termed *separatrices*. As indicated by their name, they delineate regions where all trajectories have the same asymptotic (i.e. long-term) behavior, differing from one region to the adjacent one: separatrices are the flow "drainage divide".

This hydrodynamic analogy is besides perfectly valid. The saddle point A behaves as a pass: it gathers waters flowing along directions close to the stable eigendirection and redirect them downward along directions close to the unstable eigendirection. As visible on the phase portrait, water in general does not flow through A but short-circuits the saddle, along trajectories crossing H_0 or V_0. Point B corresponds to a sink gathering all waters passing close by. Its "drainage basin" is termed its *basin of attraction*; This basin comprised all initial conditions that experience the influence of B and whose trajectories converge to B. It can be seen straightforwardly on the phase portrait that the basin of attraction of B is the region situated above the separatrices S_2 and S_4. All trajectories coming from this domain reach B more or less directly. This statement can be specified using the fact that each trajectory confines the other ones: S_1, $H_1 \cap \{x < 1\}$, $H_1 \cap \{x > 1\}$, $V_1 \cap \{y < 1\}$ and $V_1 \cap \{y > 1\}$ thus divide up the attraction basin of B in five zones conserved in the course of evolution (they are said to be *globally invariant*). Any trajectory coming from one of these zones (above S_2 and S_4) stays in it forever, whether time flows toward $t = -\infty$ or toward $t = +\infty$. On the contrary, trajectories starting below separatrices S_2 and S_4 are totally insensitive to the presence of B: this fixed point is only *locally* attractive, within its attraction basin: we here see the limit of the linear stability analysis. Below S_2 and S_4, trajectories are sensitive to the influence, first attractive then repulsive, of the saddle point A.

If trajectories are travelled in the reverse direction of time, toward $t \to -\infty$, stability is reversed: stable behaviors turn into unstable ones and vice versa, since e^{-at} (with $a > 0$) converges to 0 for $t \to +\infty$ but goes to infinity for $t \to -\infty$. A is still a saddle point (stable directions becoming the unstable ones and vice versa) but B becomes an unstable node when time is travelled backwards. According to the regions of the phase space in which they are at a given time, trajectories come:

- from $(+\infty, +\infty)$ if they pass in the first quadrant around B,
- from $(-\infty, 1)$ if they pass in the second quadrant around B,
- from $(1, \infty)$ if they pass in the fourth quadrant around B,
- from $(-\infty, 1)$ if they pass in the third quadrant around B above S_1 and S_3,

• from $(1, -\infty)$ if they pass in the third quadrant around B below S_1 and S_3.

We have thus obtained at low cost a complete knowledge of the asymptotic behavior $t \to \pm\infty$ of any trajectory. As soon as a point of the trajectory is given (one is enough), it is possible to assess which regions are visited, among those delineated by H_0, V_0 and by the particular trajectories H_1, V_1 and $(S_j)_{j=1...4}$. No integration, i.e. no explicit determination of the solutions, has been necessary. The method thus applies to any dynamical systems, whatever the complexity of the functions of (x, y) involved in the right-hand-side of $\dot{x}$ and $\dot{y}$.

This example has illustrated the notions of "basin of attraction" and "separatrix". The method of analysis extends to dimensions higher than 2; however, the drawing is less convenient, and there is a noticeable difference: trajectories now do not constitute impassable boundaries partitioning the phase space. It is less often possible to delineate invariant regions of the phase space, from which trajectories neither escape nor enter. The notion of separatrix can be extended: If A is a saddle point, it can be shown that there exists a surface made of trajectories, hence globally invariant upon evolution. This surface is tangent in A to the stable vector space; it is termed the *stable manifold* of the fixed point A. Another invariant surface exists, that is tangent in A to the unstable vector space; it is thus termed the *unstable manifold* of the fixed point A.

To conclude this paragraph, we will mention the interesting special case where the vector field V ruling the evolution is derived from a potential U, according to $V = -\text{grad}\, U$. Each component (6.1) then writes: $\dot{x}_j = -(\partial U/\partial x_j)$. Fixed points are the extrema of the potential, and the trajectories are the field lines, orthogonal in each point to the equipotential lines. In the hydrographic analogy, $U(x)$ is the altitude at point x and the landscape is described by $(x, U(x))$; trajectories followed by streaming rain water are then the steepest-descent lines.

6.2.5 Application: Anharmonic Oscillators

To illustrate the interest of the method of the phase portrait, we consider again the example of a particle in a potential well $U(x)$ in one dimension. We already encountered this problem in Chap. 1, where we used standard methods of classical mechanics, that is, solving the equation of motion $m\ddot{x} + U'(x) = 0$ to obtain the position $x(t)$ of the particle at any time $t > 0$. We met it again in Chap. 3, where we described the quantum phenomenon (tunnel effect) correcting the classical solution when scales of the system are so small that its quantum nature cannot be ignored. We will consider it now as a dynamical system in the plane (x, v). By a proper choice of the time unit, it reduces to the case where $m = 1$. In the absence of friction, the dynamical system is written:

$$\begin{cases} \dot{x} = v \\ \dot{v} = -U'(x). \end{cases} \tag{6.21}$$

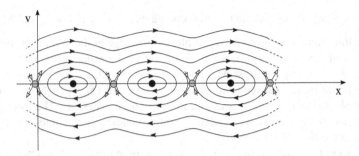

Fig. 6.2 Phase portrait of an anharmonic oscillator in a periodic potential $U(x) = U(x + 2\pi)$, for instance $U(x) = (g/l)(1 - \cos x)$ for a pendulum. The abscissa is the angle x between the pendulum and the vertical and the ordinate is the angular velocity $\omega = \dot{x}$. The scheme clearly evidences the difference between *stable equilibrium states* $x = 2n\pi$ (n being an integer) and *unstable equilibrium states* (saddle points) $x = (2n + 1)\pi$, as well as the difference between free states, associated with trajectories going from $-\infty$ to ∞, and *bound states* associated with bounded trajectories (for $E < U_{\max}$)

Conservation of the total energy, valid in the absence of friction, yields the equation of the trajectories in the implicit form $U(x) + v^2/2 = E$. It is then possible to draw the disjoint curves, parameterized by the total energy E, forming the phase portrait. Let $U_{\min} > -\infty$ be the minimum potential energy (the bottom of the well) and $U_{\max} \leq +\infty$ the maximum potential energy (energy barrier, possibly infinite), so that $U_{\min} \leq U(x) \leq U_{\max}$. Both graphical and mathematical analyses evidence two types of trajectories[8] (see Fig. 6.2):

• If $E > U_{\max}$: v does not vanish hence its sign is constant; the motion is monotonous, with a velocity $|v| \geq \sqrt{2(E - U_{\max})}$. Such a solution is termed a *free state*. Trajectories are entirely located in one of the half-planes $\{v > 0\}$ and $\{v < 0\}$ and link $x = -\infty$ to $x = +\infty$ (if v is positive, x increases from $-\infty$ to $+\infty$). Particles feel the influence of the potential, which reflects in the deflected shape of their trajectories. Acceleration $\dot{v}$ and $U'(x)$ have opposite signs, so that particles slow down when climbing the barrier, whereas they accelerate when they go down to the bottom of the well, exactly as does a ball thrown with a large enough velocity on a horizontal corrugated-iron plate. However, the influence of the potential is not strong enough to capture the particles, and the trajectories are unbounded.

• If $E < U_{\max}$: v vanishes before the particle passes the barrier[9] $U_{\max}$: it thus remains trapped in the potential well. This is also what happens for a ball on a corrugated-iron plate, when its energy is not large enough to escape a groove of the plate. The solution

---

[8]If $E = U_{\max}$, the particle comes close to the unstable state $x_{\max}$, such that $U(x_{\max}) = U_{\max}$, however with a velocity tending to 0; the particle will need an infinite time before it actually reaches $x_{\max}$, and it will never go beyond. In any case, this situation has no practical interest because it is destroyed by the slightest perturbation, which turns it in one of the situations $E < U_{\max}$ or $E > U_{\max}$; in contrast, those latter situations are robust with respect to a small enough change in E.

[9]Strictly, one should distinguish the height $U_{\max}^+$ of the barrier on the right of the well $x_{\min}$ and the height $U_{\max}^+$ of the barrier on the left of the well $x_{\min}$, and consider the sign of the initial velocity.

is then periodic and corresponds to the closed trajectories visible on the phase portrait (however, the expression of the solution is in general more complicated than a mere sine). Fixed points are the points $(x_j^*, 0)$ such that $U'(x_j^*) = 0$. The stability matrix in $(x_j^*, 0)$ is

$$M(x_j^*) = \begin{pmatrix} 0 & 1 \\ -U''(x_j^*) & 0 \end{pmatrix}. \tag{6.22}$$

Its eigenvalues satisfy $\lambda^2 + U''(x_j^*) = 0$. Two typical situations can be distinguished (for simplicity, we now omit the label j of the considered fixed point):

• If $U''(x^*) > 0$: the eigenvalues are purely imaginary $\lambda^\pm = \pm i\omega$ with $\omega^2 = U''(x^*)$. The positive sign of $U''(x^*)$ means that x^* is a well of U, hence a stable equilibrium. If the particle energy is weak enough so that it gets trapped in the well, its motion is periodic. Its period is $2\pi/\omega$ in the harmonic approximation.[10] In a more realistic model in which friction is no longer ignored, we will now show that such a fixed point becomes a stable spiral point (it is still a stable equilibrium state). Friction is taken into account through a term $-Rv$ added to the restoring force $-U'(x)$. This does not affect the position of the fixed points: friction does not influence the equilibrium states.[11] Stability matrix becomes:

$$M(x_j^*) = \begin{pmatrix} 0 & 1 \\ -U''(x_j^*) & -R \end{pmatrix}. \tag{6.23}$$

Eigenvalues are now $\lambda^\pm = -R/2 \pm (i/2)\sqrt{4\omega^2 - R}$. The fixed point is a stable spiral point as soon as friction is weak ($R < 4\omega^2$). For strong friction ($R > 4\omega^2$), eigenvalues are real and the fixed point is a stable node. Situation where $R = 4\omega^2$ corresponds to the transition between a pseudo-periodic regime (oscillatory solution whose amplitude is exponentially damped as $e^{-Rt/2}$, with a constant angular velocity $\Omega = \sqrt{\omega^2 - R/4}$, i.e. a pseudo-period $2\pi/\Omega$) to an aperiodic regime with a purely exponential damping with no oscillations.

• If $U''(x^*) < 0$: there is two real eigenvalues $\pm\lambda$ of opposite signs, with $\lambda = \sqrt{-U''(x^*)} > 0$. The fixed point is thus a saddle point and x^* lies at the top of a potential barrier. Direction $(1, \lambda)$ is unstable and direction $(1, -\lambda)$ is stable. The case $U''(x^*) = 0$ corresponds to a bifurcation where the stable equilibrium state becomes unstable.

The visualization of the dynamics by means of its phase portrait in the plane (x, v) leads to the following observations:

---

[10]In the case of small oscillations around x^*, the force $-U'(x)$ may be reduced to its linear approximation $U''(x^*)(x - x^*)$, which amounts to replace $U(x)$ by its harmonic approximation $U(x) \approx U(x^*) + U''(x^*)(x - x^*)^2/2$. Solutions of the linearized evolution are of the form $A\cos(\omega t + \varphi)$ where the constants A and φ are determined by the particle position and velocity at the initial time.

[11]However, this statement is wrong in the special case where U' vanishes on an interval. The state in which the system actually stops, among all the possible equilibrium states, would then depend on the system history and the friction it has experienced.

- The partition between free states and bound states is obvious, directly related to the (un)boundedness of the trajectory;
- Whatever the nature of trajectory (bounded or not), $|v|$ is maximal when the particle passes in a minimum of U. As soon as the particle is still moving, stable equilibrium states are those where it spends the least time! The only way to detect them experimentally is to add noticeable friction and wait for stabilization.
- $|v|$ is minimal when the system passes in a maximum of U (unstable state). The particle slows down when approaching such a saddle fixed point (think to a swing in the upside-down position) and spends there a lot of time. If its energy is sufficient, the particle manages to cross the fixed point and then accelerates away from it along the unstable direction.
- The phase portrait is modified in the presence of friction, that dissipates energy. There is no longer free states, and the total energy decreases along the time towards it minimal value U_{min}; the system is then stuck in a fixed point $(x^*, 0)$ where x^* is a minimum of U.
- If $U(x)$ is periodic and continuous, then its maxima x^*_{2j+1} alternate with its minima x^*_{2j}: $U''(x^*_{2j}) > 0$ and $U''(x^*_{2j+1}) < 0$ (letting apart the untypical case where $U''(x^*) = 0$). This alternation is the situation observed for a pendulum with x the angle between the pendulum and the downward vertical, and $U(x) = (g/l)(1 - \cos x)$. A real pendulum experiences friction. If the energy dissipation is weak enough $x^*_{2j} = 2\pi j$ are stable nodes. The system evolves toward stabilization in one of the fixed points x^*_{2j} (the larger the initial energy $E(t = 0)$, the larger $|j|$). If the pendulum is set into motion with a high initial velocity, the number j of full turns performed before the oscillatory regime starts is given by the integer part of $x(t)/2\pi$ for t tending to infinity (this integer quantity stabilizes at finite t), and it increases with the initial velocity. The same phenomenon is observed with a ball (a real one, thus experiencing friction) on a plate of corrugated iron: the number $j - 1$ of grooves that the ball crosses before being trapped in the j-th one, increases with the initial velocity.

6.2.6 The Origin of Bifurcations

An indirect, however important consequence of the flow linearization theorem, Sect. 6.2.2, is to evidence the singular character of the dynamics for which $DV(x^*)$ is not invertible. Nonlinear terms (that is, the higher-order terms in the Taylor expansion of $V(x)$ in x^*) then play an essential role. In finite dimension, non invertibility of $DV(x^*)$ corresponds to the case where at least one eigenvalue vanishes. The passage through such a situation when a parameter of V is varied may appear as an isolated and marginal event, hence far less interesting than regular behaviors far more currently observed. In fact, it is the contrary: this passage is a transition point, corresponding to a *qualitative* change of the dynamics. It is accordingly termed a *bifurcation*.

We will see in Sect. 6.3.3 that several types of bifurcations are encountered, for instance the bifurcation observed when the real part of one or several eigenvalues of $DV(x^*)$ vanishes. These bifurcations correspond to a change in the stability properties of the fixed point x^*. The reader has surely noticed that the cases let apart in the classification of the different fixed points (Sect. 6.2.3) precisely correspond to bifurcations. In Sect. 6.3 we will show the interest of the notion of bifurcation. It is so important that the study of a dynamical system depending on one or several control parameters starts with (and often reduces to) the identification of bifurcations encountered when these parameters vary.

6.3 Attractors, Bifurcations and Normal Forms

6.3.1 Attractors

We have seen the role of fixed points in the qualitative properties of the flow, in particular at long time $t \to \infty$. A *stable fixed point* corresponds to an equilibrium state in which the system S will stabilize as soon as its initial state[12] belongs to the basin of attraction of the fixed point. At a fixed point, there is no longer temporal evolution of the system state. It may happen that S stabilizes in a stationary regime more complex than a fixed point. The simplest case is a *cycle*, that is, a periodic regime. Once this regime is reached (at long enough time), the time dependence is perfectly known, up to a phase factor. This phase factor $e^{i\varphi}$ in $x(t) = e^{i\varphi}e^{i\omega t}$ simply corresponds to a shift $\Delta t = \varphi/\omega$ of the time origin. Notions of stability and basin of attraction extend to cycles. Let us consider an example.

Example 7

$$\begin{cases} \dot{x} = ax(b - x^2 - y^2) - \omega y \\ \dot{y} = ay(b - x^2 - y^2) + \omega x \end{cases} \tag{6.24}$$

with $b > 0$, $a \neq 0$ and $\omega \neq 0$. Passing to polar coordinates yields:

$$\begin{cases} \dot{r} = ar(b - r^2) \equiv v(r) \\ \dot{\theta} = \omega. \end{cases} \tag{6.25}$$

A first fixed point is $r = 0$, that is, $(x = 0, y = 0)$. Since $b > 0$, the one-dimensional dynamical system $\dot{r} = v(r)$ has another fixed point $r = \sqrt{b}$. Integration of the second equation is straightforward: $\theta(t) = \theta_0 + \omega t$. Back in Cartesian coordinates, we obtain the periodic solution (a cycle): $x(t) = \sqrt{b}\cos(\omega t + \theta_0)$ and $y(t) = \sqrt{b}\sin(\omega t + \theta_0)$, whose graph is the circle $r = \sqrt{b}$. The integration constant θ_0 is the above-mentioned phase. Computation of $v'(0) = ab$ and $v'(\sqrt{b}) = -2ab$ yields:

---

[12]Note that this initial state may be far different from the fixed point: often the basin of attraction extends up to infinity in some directions, and it may even sometimes coincide with the whole phase space.

- if $a < 0$, the fixed point 0 is stable, with basin of attraction the region $r < \sqrt{b}$, whereas the cycle is unstable. A trajectory originating in the vicinity of the cycle either tends to 0 if it starts inside the circle $r = \sqrt{b}$, or escapes to infinity if it starts outside this circle.
- if $a > 0$, the fixed point 0 is unstable. The cycle is stable and its basin of attraction is the whole plane deprived of 0. As soon as the initial point differs from 0, whatever far it could be, its trajectory wraps up around the circle $r = \sqrt{b}$.

In both cases, trajectories are spirals, turning in the direct sense if $\omega > 0$ and crossing periodically each ray with a period equal to $2\pi/\omega$. Continuing the inventory, *asymptotic motions even more complex than cycles* can be found:

- quasi-periodic motions: $x(t) = F(\omega_1 t, \ldots, \omega_n t)$, where the frequencies $\omega_1 \ldots \omega_n$ are incommensurable, that is, their ratio is irrational (see Example 2, Sect. 6.1.5),
- the case where the motion remains confined in some region of the phase space but does not present any kind of regularity: it becomes impossible to predict the evolution, i.e. the future states of S. This kind of asymptotic motion is associated with the notion of *deterministic chaos*, that will be detailed below (Sect. 6.5.2).

To gather these various asymptotic regimes, the notion of *attractor* has been introduced: it is a part $\mathcal{A}$ of $\mathcal{X}$ which is

(i) invariant: $\forall t > 0$, $f_t(\mathcal{A}) \subset \mathcal{A}$; in this respect, an asymptotic regime is stationary;
(ii) stable: trajectories starting in a neighborhood $\mathcal{U}$ of $\mathcal{A}$ will come arbitrarily close to $\mathcal{A}$. The maximal neighborhood $\mathcal{B}$ is the basin of attraction of $\mathcal{A}$. Several variants of this stability property are encountered, depending on whether trajectories come close to $\mathcal{A}$ uniformly: $\lim_{t\to\infty} f_t(\mathcal{U}) = \mathcal{A}$, or not, the convergence speed being then different for different trajectories;
(iii) indecomposable: there does not exist any strict subset of $\mathcal{A}$ satisfying the same invariance and stability properties;
(iv) maximal: there does not exist any embedding region satisfying the same invariance and stability properties, except $\mathcal{A}$ itself.

6.3.2 Conservative Versus Dissipative Systems

It is important to distinguish conservative and dissipative systems. In conservative systems, illustrated in Example 2 (Sect. 6.1.5), the volume element $d^d q d^d p$ is preserved during the evolution (Liouville theorem): there is no reduction of the accessible region of the phase space and the support of the asymptotic dynamics is the whole phase space. This conservative character is also expressed $\operatorname{div} V = 0$ (where div is the divergence operator). This condition has been encountered in hydrodynamics, Sect. 1.5.4, where it expresses flow incompressibility.[13] If $\operatorname{div} V < 0$, phase space

---

[13] However the flow is here abstract, in the phase space, whereas it is composed of real water, flowing in the real space, in hydrodynamics.

volumes are contracted by the evolution: the dynamical system is then said to be *dissipative*.

The notion of attractor is relevant only for dissipative systems: given an "appropriate" region U at time $t = 0$, we observe the contraction of U during the evolution, up to its stabilization for $t \to \infty$ on an ensemble A which is nothing but the attractor of the dynamical system. Here "appropriate" means that U has to be strictly included in the influence range of A, what is termed its basin of attraction. As in the case of a fixed point, trajectories tend to A but never reach it unless they already start in A. When the attractor is a non trivial set, it is interesting—and in fact almost sufficient—to study the dynamics restricted to the attractor.

6.3.3 The Different Types of Bifurcations

We are now in position to define precisely the notion of *bifurcation*. It is pertinent to dissipative systems, where the knowledge of the attractor is sufficient to describe the asymptotic dynamics.[14] This notion involves a parameterized family $V_\mu(x)$ of vector fields each defining a dynamical system, and the corresponding family $(A_\mu)_\mu$ of attractors describing the regime observed after stabilization. A bifurcation is a *qualitative change of the attractor A_μ* when μ crosses a value μ_c. The real number μ is termed the *control parameter* and μ_c the bifurcation value. This qualitative change can be visualized by plotting the *bifurcation diagram* representing the attractor A_μ as a function of μ (in abscissa).

Let us describe in more details the various bifurcations that may affect a fixed point x_μ^* stable for $\mu < \mu_c$. In $\mu = \mu_c$, a first possible event is the coalescence of two complex conjugate eigenvalues $\lambda^\pm = \alpha \pm i\sqrt{|\beta|}$, i.e. $\beta_{\mu_c} = 0$; they become real and distinct for $\mu > \mu_c$. The nature of the fixed point changes, from a spiral node (stable if $\alpha < 0$) to a node (with no change of stability as long as β is small enough). This qualitative change is a peculiar bifurcation since it is not associated with a change in the stability properties of the fixed point.

Strictly, an event corresponding to a true bifurcation arises when one or more eigenvalues of $DV_\mu(x_\mu^*)$ cross the imaginary axis $\Re(\lambda) = 0$. The linear approximation of the flow (Sect. 6.2.2) around x_μ^* is no longer valid and the stability properties of the fixed point change. The attractor, being by definition stable, is no longer x_μ^* if $\mu > \mu_c$. Several behaviors are typically observed (the normalization is chosen so that the bifurcation occurs for $\mu_c = 0$ with $x_c^* = 0$).

- *Saddle-node bifurcation*: $\dot{x} = \mu + x^2$. The stable fixed point $-\sqrt{-\mu}$ merges with an unstable fixed point $(\sqrt{-\mu})$ and both disappear (there is no fixed points for $\mu > 0$). We will see again this bifurcation in Sect. 6.6.2.

---

[14]Conservative systems are a different case, which will be treated in Sect. 6.6.4, with the presentation of KAM theorem; this theorem describes a behavior that can be interpreted as the analogue of a bifurcation for Hamiltonian systems.

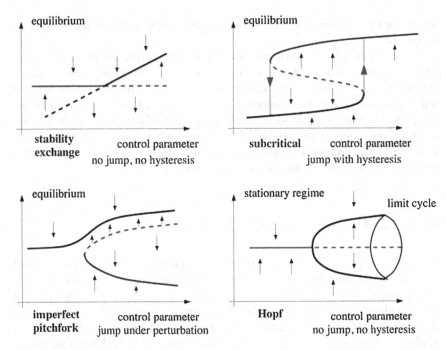

Fig. 6.3 Four examples of bifurcations

- *Transcritical bifurcation*: $\dot{x} = \mu x - x^2$. The fixed point $x^* = 0$, stable for $\mu < 0$, is replaced by $x_\mu^* = \mu$, stable if $\mu > 0$. This corresponds to an exchange of stability between the two fixed points, which both exist for any value of μ.
- *Pitchfork bifurcation*: $\dot{x} = \mu x - x^3$. The fixed point $x^* = 0$, stable for $\mu < 0$, looses its stability (but still exists for $\mu > 0$). The attractor becomes a pair of stable fixed points that appears in $\mu = 0$.
- *Hopf bifurcation*: This bifurcation exists only when the phase space dimension is larger or equal to 2. The typical example is $\dot{r} = \mu r - r^3, \dot{\theta} = 1$ (in the plane). The fixed point 0, stable if $\mu < 0$, is replaced by a stable cycle $r = \sqrt{\mu}$ for $\mu > 0$ (0 is still a fixed point but it is now unstable). We will come back to this situation in Sect. 6.6.3.

The reader is encouraged to draw the associated bifurcation diagrams. Figure 6.3. presents two of them (transcritical and Hopf bifurcations) and two additional ones (subcritical and imperfect pitchfork bifurcations). Pitchfork and saddle-node bifurcations will be encountered again in the next sections (Sect. 6.6.1, Fig. 6.6 and Sect. 6.6.2, Fig. 6.7). A bifurcation resembles a phase transition of the dynamics, which passes from one stationary regime to another, different one. As in the case of standard phase transitions, the transition can be continuous, for instance if a fixed point unfolds into a cycle at the same place in the phase space. But the transition can also be discontinuous, for instance if a fixed point becomes unstable and another stable fixed point appears at a different location in the phase space. We will describe in Sect. 6.6 the typical sequences of bifurcations leading to a seemingly erratic and unpredictable evolution, called chaos.

6.3.4 Normal Forms and Structural Stability

A dynamical system is said to be *structurally stable* if its properties, typically its long-term behavior, do not display any qualitative change when the function V involved in the evolution law (6.1) is slightly modified. Also termed *robustness*, this property differs form the previously encountered notion of stability (e.g. of a fixed point). It is an essential property insofar as it means that lacking a precise knowledge of the evolution law and ignoring some "small" influences would not spoil the conclusions of the study. Of course, what is a "small" influence has to be quantified. More generally, this notion amounts to classify the perturbations of the evolution law (6.1) according to whether they would affect or not the asymptotic behavior. This question, addressed by Morse and Weyl, has been fully solved by René Thom with the so-called *catastrophe theory* he established and the notion of *normal form*.

Let us consider a specific dynamical phenomenon, for instance a bifurcation of a given type. A *normal form* is the simplest possible function V (currently a polynomial) which can generate the phenomenon: it captures the core of the phenomenon, whence its interest. All dynamical systems displaying the same phenomenon will be equivalent to the normal form. Actually we already encountered normal forms: the examples associated above to the typical bifurcations are precisely their normal forms. A dynamical system V_0 is then structurally unstable if a perturbation changes the normal form to which it is associated: a minute modification $V_0 \rightarrow V_0 + \delta V$ puts the system in another class hence dramatically modifies its asymptotic properties.

6.4 Discrete-Time Dynamical Systems

6.4.1 Discrete-Time Evolution Equations

An important chapter of dynamical systems theory concerns evolutions where time is discrete, labelled by integers $n \in N$. It is possible to generate sequences $[x_n]_{n \in N}$ using a recursion:

$$x_{n+1} = g(x_n), \tag{6.26}$$

where g is a map from the phase space $\mathcal{X}$ into itself. Such an evolution equation is termed a *discrete dynamical system*. As in the case of continuous time, the evolution has no memory, in the sense that the knowledge of the system state at a given time fully determines its subsequent evolution. Practically, the model is relevant only if g does not depend on time (autonomous system), what will be assumed henceforth. We will present below (Sect. 6.4.4) the method of *Poincaré section* to reduce certain continuous dynamical systems to discrete analogues.

Example 8

A typical example of discrete dynamical system is provided by the logistic map $x \mapsto ax(1 - x)$ in $[0, 1]$. An equivalent form is $x \mapsto 1 - \mu x^2$ in $[-1, 1]$. It is a model of population dynamics. It describes a situation where the population cannot overwhelm a certain number N_{max} of individuals due to resource or space limitation. Each time step corresponds to a generation. N_n being the number of individuals at the n-th generation, we denote $x_n = N_n/N_{max} \in [0, 1]$. If there are few individuals ($x \ll 1$), they will develop without feeling the constraint and reproduce at a rate $a > 1$. On the contrary, if there is a lot of individuals ($N \approx N_{max}$, $x \approx 1$), the saturation of the available space or exhaustion of the available food limits the growth and the effective reproduction rate becomes $a(1 - x) \approx 0$. Merging the two regimes yields the dynamical system $x_{n+1} = ax_n(1 - x_n)$. We will see below (Sect. 6.6.1) how strongly its behavior differs from that of the continuous system presented in Example 5: the similarity of their right-hand-side is misleading, and does not at all means that the former is the discrete version of the latter. Discretization of a continuous-time dynamical system is less straightforward (Sect. 6.4.4).

6.4.2 Linear Stability Analysis

The notions introduced for systems in continuous-time can be transposed with only technical modifications which, although being important, do not change the physical interpretation. The trajectory of x_0 is the sequence $[x_n = g^n(x_0)]_{n \geq 0}$. Two trajectories are disjoint or included one into another. Here arises a marked difference with the continuous case, where a mild smoothness condition on V ensures that trajectories are defined from $t = -\infty$ to $t = +\infty$. In the discrete-time case, it is possible to define trajectories with $n \in \mathbf{Z}$ only if g *is invertible*. Indeed, if g is not invertible, the trajectories of two points having the same image upon the action of g (or one of its iterates g^n) merge. The fixed-point equation now writes $g(x^*) = x^*$. Linear stability analysis relies on the Taylor expansion at first order of $g(x) - x^*$, which leads to discuss the position of the eigenvalues of $Dg(x^*)$ *with respect to the unit circle*. An eigendirection (that is, an eigenvector of $Dg(x^*)$) is stable if and only if the associated eigenvalue Λ has a modulus $|\Lambda| < 1$; it is unstable if $|\Lambda| > 1$. A flow straightening theorem allows to identify the behavior $n \to \infty$ of the trajectory $(x_n)_{n \geq 0}$ with that of the sequence $(y_n)_{n \geq 0}$ where $y_0 = x_0$ and

$$y_n = x^* + [Dg(x^*)]^n(x_0 - x^*) \tag{6.27}$$

if $Dg(x^*)$ has no eigenvalue of modulus 1. If $Dg(x^*)$ is diagonalizable,[15] with eigenvalues $(\Lambda_j)_{j \geq 1}$ ranked such that $|\Lambda_1| \geq |\Lambda_2| \geq \ldots$, $x_0 - x^*$ is decomposed on the eigenvectors basis, which yields:

---

[15]As in the case of continuous time, if $Dg(x^*)$ is not diagonalizable and possesses a Jordan block $k \times k$ associated with the eigenvalue Λ, terms $n\Lambda^n, \ldots, n^{k-1}\Lambda^n$ appear. They do not affect the stability properties of the fixed point.

$$y_n = x^* + \sum_{j \geq 1} \Lambda_j^n c_j(x_0). \tag{6.28}$$

The asymptotic behavior is ruled by $|\Lambda_{j_0}|^n$ if c_{j_0} is the first non vanishing coefficient in the expansion. The various types of fixed points described in Sect. 6.2.3 have discrete analogues, that the reader may list as an exercise.

6.4.3 Attractors and Bifurcations

Some results are specific to discrete dynamical systems. A *n-cycle* is a periodic trajectory $(x_0, \ldots, x_{n-1})$ with period n; each point of this cycle has to be a fixed point of g^n (in particular $x_n = x_0$). Stability of this cycle is that of any of these fixed points with respect to the evolution law g^n, hence it is determined by the eigenvalues of $Dg^n(x_0) = Dg(x_{n-1}) \cdots Dg(x_0)$. Stability matrices $Dg^n(x_j)$ of the other fixed points are in general different, however they have the same eigenvalues.[16] A n-cycle is thus stable if its elements are stable fixed points of g^n.

Bifurcations correspond to the passage of one or more eigenvalues across the unit circle. Passage through $\Lambda_c = 1$ is the discrete version of the *saddle-node bifurcation*. Passage through $\Lambda_c = -1$ is the discrete version of the *pitchfork bifurcation*; corresponding to the transition from a fixed point to a stable 2-cycle, or from a 2^n-cycle to a 2^{n+1}-cycle if it is observed on g^n. For this reason, it is also termed *period-doubling bifurcation*. An *Hopf bifurcation* is observed when a pair of complex conjugate eigenvalues crosses the unit circle.

Attractors can be more complex than cycles. As an attractor $\mathcal{A}$ is by definition invariant ($g(\mathcal{A}) \subset \mathcal{A}$), it contains the trajectory of each of its points. It is then often enough to consider the dynamical system $g_{|\mathcal{A}}$ restricted to the attractor. *This reduced system contains complete information on the asymptotic behavior.*

6.4.4 Discretization by Poincaré Sections

The most often encountered method to reduce the study of a dynamical system in continuous time to a dynamical system in discrete time is the *Poincaré section* method. It applies when the continuous flow f is approximately periodic in the neighborhood of a point x_0. One then chooses an hyper-surface Σ containing x_0 and crosswise to the flow termed a *Poincaré section*. If possible, this surface is defined through an observable physical criterion: vanishing of a phase, crossing of a special value for a given observable...The assumption of almost-periodicity implies that a trajectory starting in a point $x \in \Sigma$ close enough to x_0 comes back in Σ in a

---

[16]The basic argument is: $\mathrm{Det}(A_1 A_2 - \lambda \mathbf{1}) = \mathrm{Det}(A_2 A_1 - \lambda \mathbf{1})$ hence $A_1 A_2$ and $A_2 A_1$ have the same eigenvalues, even if $A_1 A_2 \neq A_2 A_1$.

point $f(s_x, x) \equiv g(x)$. The moment s_x is termed *the first-return time*; in general it varies when x varies. The map $g(x)$, termed the *first-return map*, is defined in a neighborhood of x_0 in Σ. This map g generates an autonomous discrete-time dynamical system whose merit is to be adapted to the continuous evolution: the time intervals s_x involved in the expression of g are intrinsic, i.e. prescribed by the dynamics and not by an external observer. Another advantage of g is to be defined in Σ and with values in Σ and not[17] in $\mathcal{X}$: the phase space dimension, that is, the number of degrees of freedom to be handled, has been reduced by at least one unit.

If the discrete dynamical system has been obtained using the Poincaré section method, then a fixed point x^* corresponds to a cycle of the continuous dynamical system, with period $T = s(x^*)$. The eigenvalues $(\lambda_j)_{j \geq 1}$ of the continuous dynamical system[18] describing the stability properties of a cycle are simply related to the eigenvalues $(\Lambda_j)_{j \geq 1}$ of $Dg(x^*)$ according to

$$\Lambda_j = e^{T\lambda_j}, \tag{6.29}$$

which demonstrates the equivalence of the stability criteria:

$$\left[|\Lambda_j| < 1 \right] \Longleftrightarrow \left[\Re(\lambda_j) < 0 \right] \Longleftrightarrow \text{ linear stability.} \tag{6.30}$$

It is thus legitimate to investigate the stability of the fixed point x^* of g instead of the stability of the cycle: the conclusions will be identical.

To illustrate this procedure, let us come back to the above Example 7 (Sect. 6.3.1) with $b = 1$. The continuous flow possesses a cycle $\{r = 1\}$. We may choose the ray $\Sigma = \{\theta = 0, r > 0\}$ as a Poincaré section (or any other ray). This half-line Σ is *crosswise to the flow* insofar as trajectories cross it with a non-vanishing angle ($\dot{\theta} = \omega > 0$). Moreover, if a trajectory starts in $x_0 \in \Sigma$, it crosses again Σ: first-return times are all identical, equal to the period $T = 2\pi/\omega$ of the cycle. The Poincaré map is given by $f_T(r)$ where f_t is the continuous flow generated by $\dot{r} = v(r) = ar(1 - r^2)$. We recover the general statement: the discretization has skipped one degree of freedom, here θ. The stability of the cycle can be determined by investigating the stability of the fixed point $r^* = 1$ (intersection of the cycle and Σ) upon the action of g, or as we have done before, upon the action of the flow f_t. This example is however peculiar insofar as radial and angular components are decoupled, which makes the discretization quite trivial, except for illustration purposes.

---

[17]On the contrary, the "naive" discretization $g(x) = f_{\Delta t}(x)$ where the time step Δt is identical for all trajectories and all steps yields a discrete dynamical system in $\mathcal{X}$.

[18]The continuous dynamical system has an additional eigenvalue $\lambda_0 = 0$ corresponding to the invariance upon a translation by an integral number of turns along the cycle. The reader can easily check this statement on Example 7.

6.5 Lyapunov Exponents and Deterministic Chaos

6.5.1 Lyapunov Exponents

Differentiating (6.1) yields an equation describing the time evolution of the separation[19] $\Delta x(t, x_0, \Delta x_0)$ observed at time t between the trajectories starting in x_0 and $x_0 + \Delta x_0$:

$$\partial_t \Delta x(t) = DV[x(t)]\dot{\Delta}x(t). \tag{6.31}$$

This equation is very interesting since it describes how two initially close trajectories move apart or come close one to another along time. In the neighborhood of a fixed point x^*, $x(t)$ can be replaced by x^* in $DV[x(t)]$, which reduces (6.31) to an easily integrable linear differential equation with constant coefficients: the solution is a linear combination of exponential functions $e^{\lambda t}$ where the exponents λ are the eigenvalues of $DV(x^*)$.

Far from a fixed point, (6.31) is still linear with respect to Δx however it is more difficult to integrate this equation because the coefficients of the matrix $DV[x(t)]$ depend on time, through their dependence on the reference trajectory $x(t) = f_t(x_0)$ (which besides is not always explicitly known). A first important result of the dynamical systems theory (due to Lyapunov and Birkhoff, at the beginning of the last century) was to show that the time dependence of $\Delta x(t)$ remains *asymptotically* exponential, however with other exponents $(\gamma_j)_{j \geq 1}$ termed the *Lyapunov exponents*. In dimension 1:

$$\gamma = \lim_{T \to \infty} \int_0^T V'(x(t))\, dt. \tag{6.32}$$

γ is a global characteristic of the flow, since its expression involves the value of V' in each point $x(t)$ of the trajectory ($t \in [0, \infty[$). Equation (6.32) defines a quantity $\gamma(x_0)$ which a priori depends on the considered trajectory. A key result, obtained within the framework of *ergodic theory* (see Sect. 6.5.3) is that for a certain class of evolutions, γ does not depend on x_0 and can be calculated as

$$\gamma = \int_{\mathcal{X}} V'(x)\, \rho(x)dx, \tag{6.33}$$

where $\rho(x)$ is a density[20] preserved along the evolution.

---

[19] Strictly, like $V[x(t)]$, $\Delta x(t, x_0, \Delta x_0)$ belongs to the tangent space of $\mathcal{X}$ in $x(t) = f_t(x_0)$.

[20] Strictly, one should rather speak of the "invariant measure" m of the dynamical system and writes the infinitesimal volume element $dm(x)$. It can be written $dm(x) = \rho(x)dx$ as soon as m possesses a smooth density.

Fig. 6.4 Sensitivity to initial conditions. Two initially close trajectories ($a_0 \approx a_0'$) move apart exponentially fast. This chaotic feature is quantified by a Lyapunov exponent γ, describing the separation rate *averaged over the whole evolution*: $||a_t a_t'|| \sim e^{\gamma t}||a_0 a_0'||$ (for $t \to \infty$)

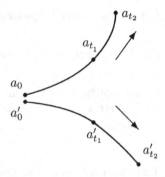

This statement can be generalized to higher dimensions, and it allows one to define and compute a sequence[21] of exponents $(\gamma_j)_{j \geq 1}$ ranked in decreasing order. The separation $\Delta x(t, x_0, \Delta x_0)$ between two trajectories initially separated by a distance Δx_0 has a complicated evolution, with for instance a non monotonous dependence on t on short time intervals. However, the long-term trend is not only very simple but also identical for all typical trajectories. One has

$$\lim_{T \to \infty} \frac{1}{T} \log \left(\frac{\Delta x(t)}{\Delta x_0} \right) = \gamma_1, \tag{6.34}$$

unless Δx_0 belongs to a special hyper-surface of $\mathcal{X}$, which happens with a null probability (for the same reason that a surface has a null volume). On this hyper-surface, the next exponent γ_2 is observed (with $\gamma_2 < \gamma_1$), or even γ_3 is observed (with $\gamma_3 < \gamma_2$) on a surface of lower dimension.

Lyapunov exponents are essential when the flow has no simple attractor. The maximal Lyapunov exponent γ_1 alone characterizes almost fully the dynamics. As soon as $\gamma_1 > 0$, initial discrepancies (fluctuations δx_0, errors in the location of x_0...) are amplified: one speaks of *sensitivity to initial conditions* (Fig. 6.4).

6.5.2 Deterministic Chaos

The chaos (of course in a mathematical sense) is a recent notion, developed about forty years ago. After a few years of debate, a consensus arose on the following characterization: a deterministic evolution is *chaotic*
(i) if it displays the above-described property of *sensitivity to initial conditions* (in other words if it possesses at least one strictly positive Lyapunov exponent),

---

[21] The exponents $(\gamma_j)_{j \geq 1}$ in general differ from the eigenvalues $(\lambda_j)_{j \geq 1}$, unless the flow as a unique stable fixed point with a basin of attraction equal to the whole phase space $\mathcal{X}$. Indeed in this case, trajectories rapidly reach a neighborhood of x^* where $DV[x(t)]$ and $DV(x^*)$ can be identified.

Fig. 6.5 Mixing mechanism
in the phase space (A_1 is the
image of A_0 and so on)

(ii) and if it exists a nonlinear mechanism ensuring the *mixing* of the trajectories and
their re-entry in a bounded domain of the phase space (Fig. 6.5).

Sometimes an extra condition is added: the evolution has to take place in a space
of *finite dimension* and even of *low dimension* (else one speaks of *turbulence*). An
imagery is the making of puff pastry or bread (Fig. 6.4). The evolution $\dot{x} = 2x$ on
the real line is not chaotic since the trajectories go to infinity in a monotonous way:
it satisfies condition (i) but not (ii). In contrast, the evolution $\dot{x} = 2x \, modulo \, 2\pi$ (on
the circle) is chaotic, it is even an emblematic example of chaos.

The notion of chaos for dissipative systems is often associated with that of strange
attractor. A *strange attractor* is first an attractor, of null volume since the dynamical
system is dissipative, however of very complex geometry. Its structure is self-similar
at all scales, whatever small: it is an instance of fractal structure (see Chap. 5). The
second property defining a strange attractor is the chaotic nature of the dynamics
restricted to this attractor (existence of a strictly positive Lyapunov exponent): the
asymptotic behavior is thus unpredictable.

The model introduced by Lorenz (Example 4, Sect. 6.1.5) to describe the convection of an incompressible fluid provided one of the first strange attractors ever
encountered: the Lorenz attractor (Fig. 6.6). Some chemical systems may present
chaotic behaviors that we can directly observe, at our scale. Another example is that
of a compass put in two magnetic fields, one fixed and the other rotating: the evolution preserves the total energy of the physical system comprising the compass and
the rotating field (phase space of dimension 4), however it is chaotic. Chaos may
thus occur in conservative systems as well, however only dissipative systems display
strange attractors.

The role of the phase space dimension d regarding chaos is best discussed using
the phase portrait visualizing the flow in the phase space. In dimension 1 or 2,
trajectories behave as impassable boundaries, which ensures that the fate of a point
is perfectly predictable, on any term. *No chaos can arise if $d \leq 2$.* If $d \geq 3$, none of
the trajectories achieves a partition of $\mathcal{X}$: two arbitrarily close points, experimentally
indistinguishable, could have on the long term totally different evolutions; in this
sense their evolution is unpredictable. The constraint $d \geq 3$ apparently disappears
for discrete dynamical systems, due to the fact that they are currently obtained as a
projection and section of continuous flows in higher dimension.

Fig. 6.6 Lorenz attractor (courtesy of H. Lemarchand). The evolution law is given in the Example 4, Sect. 6.1.5. The represented attractor is obtained for $\sigma = 10$, $r = 28$ and $b = 8/3$. It is a typical example of *strange attractor*: it has a fractal structure and the dynamics restricted to this attractor is chaotic (mixing, dilating the distances in some directions and contracting the distances in crosswise directions)

6.5.3 *Ergodic Theory*

To really understand the results brought by the theory of dynamical systems, or even to state them, it is necessary to invoke some notions of the (mathematical) *measure theory* and its subdomain specific to dynamical systems, the *ergodic theory*. This is far beyond the (physical) scope of this book, however we may briefly outline this viewpoint. The main notion is that of *invariant measure*: its a weighting[22] $\rho(x)$ of the points x of $\mathcal{X}$, which remains unchanged when each of these points x has evolved into x_t. Each point has moved along its trajectory but the global picture provided by the probability of presence ρ has not changed: fruits in the t-slice of the cake (already savored above Sect. 6.1.3) are not the same as those of the initial slice, however they lie at the same location. Studying the invariant measure provides a lot of information on the dynamics.

A key result is the *ergodic theorem of Birkhoff*. Under a certain assumption on the flow and its invariant measure, termed *ergodicity*,[23] any observable satisfies:

---

[22]To simplify, we consider the simplest case where the measure is defined by a density: $dm(x) = \rho(x)dx$. It is not always the case, and strictly all should be written using the volume element $dm(x)$.

[23]A flow, with invariant measure m, is ergodic if any invariant subset $\mathcal{B}$ of $\mathcal{X}$ ($f(\mathcal{B} \subset \mathcal{B})$ either has null measure ($m(\mathcal{B}) = 0$), or its complementary has null measure ($m(\mathcal{X} - \mathcal{B}) = 0$). This property of the pair (f, m) is best understood using the equivalent definition of ergodicity: *a typical trajectory visits an infinite number of times any region $\mathcal{B}$ of $\mathcal{X}$ provided the measure of $\mathcal{B}$ does not vanish (even if it is arbitrary small)*.

$$\bar{A}(x_0) \equiv \lim_{T \to \infty} \frac{1}{T} \int_0^T A(x_t)dt = \int_{\mathcal{X}} A(x)\rho(x)dx \equiv \langle A \rangle. \qquad (6.35)$$

The left-hand-side is the *time average* $\bar{A}(x_0)$ of A along the trajectory starting in x_0. The theorem thus states the equality between this time average and the statistical average $\langle A \rangle$ with respect to the invariant density ρ, *whatever the starting point x_0*. Consequently, the time average along a trajectory does not depend on the considered trajectory, and it can be computed through an integration over $\mathcal{X}$, termed an *ensemble average*, as soon as the invariant measure is known and ergodicity applies. We recover here one of the bases of statistical mechanics, known as the *ergodic hypothesis* and introduced by Boltzmann. This hypothesis amounts to postulate the ergodicity of the canonical distribution with respect to the microscopic dynamics. The requirement of ergodicity and the justification of macroscopic behaviors starting from microscopic evolution laws aroused heated debates at Boltzmann's time, some of them lasting up to now.

6.6 Routes to Chaos

A *route to chaos* is a specific sequence of bifurcations leading from a fully predictable evolution (for instance, possessing a stable fixed point) to a chaotic evolution. A remarkable feature, discovered in the 1980s, is that these sequences are often qualitatively identical, even if the physical systems are totally different. This is termed the *universality* of the routes to chaos.

Three typical routes have been evidenced for dissipative systems, each associated with a type of bifurcation: the period-doubling route associated with the pitchfork bifurcation, the intermittency associated with the saddle-node bifurcation, and the Ruelle-Takens associated with the Hopf bifurcation. We will end this section by describing what may happen in conservative systems (i.e. Hamiltonian systems).

6.6.1 Period Doubling and Subharmonic Cascade

This route describes the passage from a situation where the physical system S reaches an equilibrium state (that is, a stable fixed point) to a chaotic regime through a sequence of period doublings (in mathematical terms, pitchfork bifurcations, Sect. 6.3.3). As the control parameter μ is increased, the fixed point is replaced in $\mu = \mu_0$ with a cycle having a certain period T (the fixed point still exists for $\mu > \mu_0$ but it is then unstable). Then, in $\mu = \mu_1$, this cycle looses its stability and is replaced by a stable cycle of period $2T$. And so on: the stable cycle of period $2^{j-1}T$ observed for $\mu < \mu_j$ looses its stability $\mu = \mu_j$; simultaneously, it appears a stable cycle a period $2^j T$ (twice the former) which becomes the attractor for $\mu > \mu_j$.

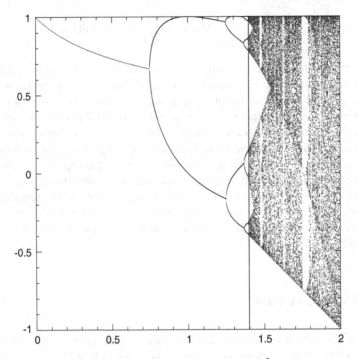

Fig. 6.7 Bifurcation diagram of the logistic map $f(x) = 1 - \mu x^2$. The parameter μ is put in abscissa and the attractor along the other axis. This plot clearly displays an accumulation of period doublings leading to chaos in $\mu_c = 1.4011550\dots$

A first remarkable feature is that period doublings succeed each other indefinitely (Fig. 6.7). The increasing sequence $(\mu_j)_{j\geq 0}$ of bifurcation values tends to a limit μ_c. In $\mu = \mu_c$, a chaotic regime appears. The second and even more striking feature is a universality property of this route. Bifurcation values $(\mu_j)_{j\geq 0}$ are specific to the considered system, as well as their limit value μ_c. In contrast, their accumulation in μ_c follows a geometric progression:

$$\lim_{j\to\infty} \frac{\mu_{j+1} - \mu_j}{\mu_{j+2} - \mu_{j+1}} = \delta \quad \Longleftrightarrow \quad \mu_c - \mu_j \sim \delta^{-j}, \qquad (6.36)$$

where δ is a universal number: $\delta = 4,66920\dots$ This means that the exponent δ is identical in all the systems where such a sequence of period doublings leading to chaos is observed: a *qualitative* similarity between asymptotic behaviors implies a *quantitative* identity.

This route is also termed *subharmonic cascade* since in the frequency space, a pitchfork bifurcation corresponds to the appearance of a peak at half the frequency (subharmonic) of the already existing one. For $\mu_j < \mu < \mu_{j+1}$, the spectrum displays peaks in $\omega = \omega_0, \omega_0/2, \ldots, \omega_0/2^j$. This route is the most acknowledged of the routes to chaos, for the following reasons:

- this route is observed in numerous experimental situations, for instance Rayleigh-Bénard convection, and experimental measures support its universality and the value of δ;
- this route is easily observed numerically, for instance in the family of discrete dynamical systems generated by the parameterized map $g_\mu(x) = 1 - \mu x^2$. The bifurcation observed in $\mu = \mu_j$ corresponds to the loss of stability of a 2^{j-1}-cycle, stable for $\mu_{j-1} < \mu < \mu_j$, jointly with the appearance of a stable 2^j-cycle which becomes the attractor for $\mu_j < \mu < \mu_{j+1}$;
- last but not least, this route is fully described and understood analytically. It is in this way that the universality of δ is proven and its value computed. It is even possible to delineate the families of dynamical systems that follow this route: they form what is called the *universality class* of the route.

6.6.2 Intermittency

The route named intermittency describes the persistence of regular and predictable phases in a globally chaotic dynamics. The key idea is that after the disappearance of a stable fixed point x_μ^* through a saddle-node bifurcation in $\mu = 0$, the dynamics remains slow in the neighborhood of x_0^*, as if it were experiencing the presence of a ghost fixed point. The typical example (actually, the normal form) is the discrete evolution:

$$x_{n+1} = g_\mu(x_n) = -\mu + x_n - A x_n^2. \tag{6.37}$$

In $\mu = 0$, the fixed points $\pm\sqrt{-\mu/A}$ observed for $\mu < 0$ (respectively stable and unstable) merge in $x_0^* = 0$ and for $\mu > 0$, there is no longer fixed points. However, $g_\mu(x) \approx x$ in the neighborhood of 0, so that the trajectory loiters a long time in this region and a regular and slow regime is observed, that roughly follows the evolution law $x_{n+1} = x_n - \mu$ as long as $A x_n \ll 1$ and $A x_n^2 \ll \mu$ (Fig. 6.8). The average duration $T(\mu)$ of these regular phases is defined as the number of steps remaining in the neighborhood of 0, averaged over all trajectories experiencing this trapping. It can be shown that this duration $T(\mu)$ depends on μ according to

$$T(\mu) \sim \frac{1}{\sqrt{\mu}} \qquad (\mu > 0). \tag{6.38}$$

Fig. 6.8 Temporal
intermittency. This sketch
explains the slow and regular
regime displayed by the
discrete evolution
$x_{n+1} = g_\mu(x_n)$ if μ is
slightly larger than the value
μ_0 associated with the
bifurcation $g'_{\mu_0}(0) = +1$

6.6.3 Ruelle-Takens Scenario

The exact statement of this route is quite technical since it requires to introduce
a topology on the space of vector fields $\{V\}$, in order to define the closeness of
two continuous dynamical systems. Its approximate formulation is written: *a continuous dynamical system experiencing three successive Hopf bifurcations typically
possesses a strange attractor* (Fig. 6.6).

Mathematical refinements that we here let apart make more precise the meaning
of "typically". Each Hopf bifurcation corresponds to the appearance of an unstable
mode.[24] The above statement can be reformulated: *the loss of stability of 3 modes
with frequencies whose pairwise ratios are irrational leads to chaos.* This result has
deeply modified the image scientists had of chaos: the previously prevailing vision,
due to Landau, required the loss of stability of an infinity of modes for the evolution
to become seemingly erratic and unpredictable. Accordingly, chaos was believed
to possibly arise only in systems with an infinity of degrees of freedom. Landau
scenario proved to be by far too restrictive: the nonlinear coupling of three modes
whose pairwise frequency ratios are irrational is sufficient to generate a strange
attractor.

6.6.4 Hamiltonian Systems and KAM Theorem

To end this section, we mention a very important result concerning Hamiltonian
systems (see Example 2, Sect. 6.1.5): the *KAM theorem*. This theorem does not
really propose a route to chaos, however it describes the effect of a non integrable
perturbation on an integrable system with Hamiltonian H_0. The control parameter is

---

[24] A *mode* is a component $c_\lambda e^{\lambda t}$ of an evolution $f(t)$. The coefficient c_λ is the amplitude of the
mode; it is related to the Fourier transform of f (if $t \in]-\infty, +\infty[$ and f integrable) of the Laplace
transform of f (if $t > 0$). The mode is stable if $\Re(\lambda) < 0$, unstable if $\Re(\lambda) > 0$ (for $t \to +\infty$).

Fig. 6.9 Loss of stability of an integrable Hamiltonian system. **a** For $\epsilon = 0$, invariant curves are the circles $x = x_0 = $ const., on which the dynamics reduces to a rotation of angle $\omega(x_0)$. **b** As soon as $\epsilon > 0$, circles with rotation number $\omega(x_0)$ break down if $\omega(x_0)$ is *rational*; they are deformed in a differentiable manner if $\omega(x_0)$ is *irrational*. **c** Chaotic zones appear but they are trapped in between the closed invariant curves still present. **d** Above a value ϵ^*, all invariant curves have disappeared. The evolution is fully chaotic, displaying sensitivity to initial conditions, and the trajectories can cross the whole phase space in the x-direction (*dotted curve*)

here the amplitude ϵ of the perturbation:

$$H_\epsilon = H_0 + \epsilon V. \tag{6.39}$$

The existence of non integrable Hamiltonian systems, generating more complex motions than quasi-periodic motions, has been evidenced by Poincaré. This result played an important role since it compelled to give up the idea of representing any motion as a combination of periodic motions (an idea that dated back to Ptolemy and is implemented in a more modern way by the action-angle coordinates presented in Example 2, Sect. 6.1.5, for an integrable system). Reflecting the names of its fathers, Kolmogorov, Arnold and Moser, the KAM theorem describes the distortion and the disappearance of the quasi-periodic solutions $q(t) = Q(\omega_1 t + \theta_1^0, \ldots, \omega_d t + \theta_d^0)$ of the initial system when ϵ increases.

The phase space is the space $\{J, \theta\}$ of action-angle coordinates, that is, $\mathbf{R}^d \times \mathbf{T}^d$ if there are d mechanical degrees of freedom, hence d actions $J_1, \ldots, J_d$ and d angles $\theta_1, \ldots, \theta_d$. In this space, solutions delineate invariant spaces, *KAM tori*,[25] that behave as impassable boundaries. It is then possible to predict the fate of a point since its trajectory remains trapped in the region where it starts (Fig. 6.9).

The gradual disappearance of these invariant sets as ϵ increases gives more space to the other trajectories. Accordingly, what can be predicted on the future location in the phase space of a system knowing its initial state decreases. The KAM theorem demonstrates that above a threshold value ϵ_c, there is no longer an invariant space partitioning the phase space, so that nothing can be predicted about the fate of a point. In this respect, the crossing $\epsilon = \epsilon_c$ corresponds to a transition toward a long-term uncontrolled dynamics. To speak of a transition to chaos, it is moreover necessary to evidence sensitivity to initial conditions (that is, the existence of at least one positive Lyapunov exponent) and a mixing mechanism.

---

[25]If $\epsilon > 0$, the tori are deformed: they are termed tori as long as the deformed set remains smoothly related (diffeomorphic) to the torus $\{J_1, \ldots, J_d\} \times \mathbf{T}^d$ observed for $\epsilon = 0$.

Let us restrict ourselves to the case of two mechanical degrees of freedom, that is, a phase space of dimension 4. A projection on the hyper-surface $H_\epsilon = $ const., of dimension 3, followed by a Poincaré section, leads to investigate a reduced discrete-time evolution on the cylinder $S^1 \times \mathbf{R}$, of the form:

$$g_\epsilon : \quad \begin{pmatrix} \theta \\ x \end{pmatrix} \longrightarrow \begin{pmatrix} \theta + \omega(x) + \epsilon \, a(\theta, x) \pmod 1 \\ x + \epsilon \, b(\theta, x) \end{pmatrix} \qquad (6.40)$$

Originating from a conservative system, g_ϵ preserves the measure of the areas on the cylinder. For $\epsilon = 0$, all the circles $x = const.$ are invariant and the circle $x = x_0$ is travelled with an angular velocity $\omega(x_0)$. The KAM theorem describes the distortion and the disappearance of these invariant circles as ϵ increases. Remarkably, *a circle is the more robust the more irrational $\omega(x_0)$ is*. This feature can be explained: the solution involves an infinite sum $\sum_{n=0}^\infty c_n [e^{2i\pi n\omega} - 1]^{-1}$. If ω is rational, $\omega = p/q$, the terms for which n is a multiple of q go to infinity. Such a term is called a *resonance*, since the divergence gives rise to the well-known physical phenomenon of resonance. These resonances raise a difficulty also called the *issue of the small denominators*. If on the contrary ω is irrational, all the terms in the infinite sum are bounded. The convergence of the sum is the better the more irrational ω is. The last circle to disappear is the circle for which $\omega(x_0) = \frac{1+\sqrt{5}}{2} \equiv \sigma$ (the golden ratio, which is the most irrational number in $[1, 2]$).

This theorem, whose complete statement is very technical, is however one of the most fascinating of the theory of dynamical systems, as it involves a domain of number theory, called Diophantine arithmetics.[26] This theory provides the mathematical tools to quantify the level of irrationality of ω, and to hierarchically rank the various values taken by ω from the rational numbers up to the most irrational one, the golden ratio σ. The KAM theorem thus relates a practical physical problem: the stability of an integrable Hamiltonian evolution with respect to non integrable perturbations, to the purest part of mathematics. It gives a high role to the celebrated number $\sigma = \frac{1+\sqrt{5}}{2}$, the *golden ratio* of the ancient greeks. This would have been the stuff of Pythagoras[27] dreams!

6.7 Conclusion

Our overview of dynamical systems is now achieved. A first interest of this viewpoint on dynamics is to take direct benefit of mathematical notions and results to answer questions addressed on the long-term behavior of a physical system. But the principal interest goes far beyond this technical advantage. By pruning the description

---

[26] Dubbed from the greek mathematician Diophantus of Alexandria (325–410).

[27] Acknowledged for a mundane theorem that was besides established earlier, Pythagoras (560–480) was also a mystical philosopher; we owe him the first results of number theory, in particular the irrational nature of $\sqrt{2}$.

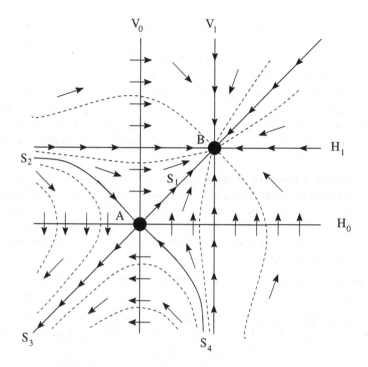

Fig. 6.10 Phase portrait of Example 6, Sect. 6.2.4. Information lies in the value $\alpha(x, y) = x(1 - y)/y(1 - x)$ of the slope of the trajectory passing through (x, y). One then determines the sign and variation of this slope in each of the phase space regions delineated by boundary trajectories termed separatrices

and focusing on the phase space geometry of the dynamics, it has pointed out the essential concepts for understanding the mechanisms ruling the evolution of complex physical systems: *asymptotic dynamics, attractors and basins of attraction, bifurcations, normal forms and classification of nonlinearities in the dynamics, Lyapunov exponents, deterministic chaos and universal routes to chaos.* In perspective, this theory has not given all its fruits: it is now explored further to describe and understand far-from-equilibrium systems and turbulence.

6.8 Problems

6.1 Non autonomous dynamical systems: Show[28] that a continuous dynamical system in $\mathcal{X}$ whose evolution law depends explicitly on time $dx/dt = V[t, x(t)]$ reduces to an autonomous dynamical system in $\mathbf{R} \times \mathcal{X}$. Same question in the discrete case: $x_{n+1} = f_n(x_n)$.

---

[28] Hint: introduce an additional variable s, with $s(t) \equiv t$.

6.2 Dynamical system with memory: Show[29] that a discrete dynamical system in $\mathcal{X}$ where the state at time n depends on $k > 1$ previous states reduces to a usual dynamical system in $\mathcal{X}^k$. What conditions are to be specified to fully prescribe a trajectory? Determine the equation for its fixed points.

6.3 The simple pendulum: In the spirit of Example 1 (Sect. 6.1.5), write the continuous dynamical system describing the motion of a simple pendulum with mass m and length l. Determine the fixed points and their stability. Draw the trajectories in the plane $\{\theta, \dot{\theta}\}$ where θ is the angle between the pendulum and the downward vertical.

6.4 The damped harmonic oscillator: Write the continuous dynamical system describing a mass m bound to the end of a vertical spring of rigidity K and immersed in a liquid of viscosity γ. Determine the fixed points, then investigate their nature and their stability, according[30] to the values of the parameters m, K and γ. Draw the trajectories in the phase space $\{z, \dot{z}\}$ where z is the mass altitude, measured with respect to the equilibrium position $z = 0$.

6.5 Chemical systems: Investigate the asymptotic behavior $t \to \infty$ of chemical systems in Example 3 (Sect. 6.1.5): $A + 2B \xrightarrow{k} 2C + D$ and $A + B \xrightarrow{k} 2A$. Investigate the dynamical system describing the chemical reaction: $A + nB \xrightarrow{k} mC$. (Discuss according to the values of integers n et m). In particular, show that if initial concentrations are positive, they remain so along time.

6.6 A hyperbolic system: Draw the phase portrait of the system:

$$\begin{cases} \dot{x} = Ax + By \\ \dot{y} = Bx + Ay \end{cases} \qquad (B > |A|) \tag{6.41}$$

6.7 The logistic map: Consider the discrete-time equation $x_{n+1} = f_\mu(x_n)$ where $f_\mu = 1 - \mu x^2$. For which values of μ does this map define a discrete-time dynamical system on $[-1, 1]$? Show that there exists a fixed point x_μ^*. For which values of μ is this fixed point stable? Show that it looses its stability at the same time that a stable 2-cycle appears. Up to which values of μ is this 2-cycle stable?

6.9 Further Reading

The bibliography on deterministic chaos is huge, even more that on dynamical systems. We cite here only a few.

Mathematical notions can be found in: Arnold V.I. (1978) *Ordinary differential equations*, the MIT Press; Arnold V.I. & Avez (1968) Ergodic problems of classical mechanics, Benjamin.

---

[29] Hint: define a k-dimensional state variable $X_n = (x_{n-k+1}, \ldots, x_{n-1}, x_n)$.
[30] Hint: a qualitative change takes place for $\gamma^2 = 4Km$.

Bifurcation theory is presented in: Ruelle D. (1989) *Elements of differentiable dynamics and bifurcation theory*, Academic Press; Demazure M. (2000) *Bifurcations and catastrophes: geometry of the solutions to nonlinear problems*, Springer.

Routes to chaos are presented in: Bergé P., Pomeau Y. & Vidal C. (1987) *Order within chaos*, Wiley VCH, and their universality properties in: Lesne A. (1998) *Renormalization methods*, Wiley & Sons.

Link to hydrodynamics is discussed in: Manneville P. (1990) *Dissipative structures and weak turbulence*, Academic Press.

A reflexion on the notion of deterministic chaos by its founders is offered in: Ruelle D. (1993) *Chance and Chaos*, Princeton University Press; Lorenz E.N. (1995) *The essence of chaos*, Taylor & Francis.

The link between chaos and statistical mechanics, with an introduction to dynamical systems and ergodicity, can be found in: Castiglione P., Falcioni M., Lesne A. & Vulpiani A. (2008) *Chaos and coarse-graining in statistical mechanics*, Cambridge University Press.

A more detailed however still introductory presentation of all the topics described in this chapter can be found in: Laguës M. & Lesne A. (2011) *Scale invariance*, Springer (Chap. 9).

Appendix A
Elementary Particles

Many decades ago, when we the authors tried to learn physics, the field of elementary particles grew out of nuclear physics and was quite separate from what is presented in the earlier chapters of this book. In recent years, however, statistical physics and the theory of elementary particles haved moved closer together, e.g., by exchanging computer experience: the statistical physics of Sect. 4.3.9 learned from particle theoreticians, and statistical models as in Fig. 5.6 served as test cases for particle computer algorithms. Thus we try to summarize here the present status of this rapidly developing area, without asserting that this must be standard knowledge. Actually we learned it from Professors Rollnik and Heinloth of Bonn University, who in 1993 as missionaries preached quark wisdom to the heathen natives at Cologne, and from reviews of Jean Potvin at Cahokia. It may be taught directly after Chap. 4; if Chap. 5 is a lunch dessert, the present chapter is some sort of five o'clock tea, and as the Boston Tea Party made clear, not everybody likes such tea.

A.1 Basic Facts and Other Lies

A.1.1 Particles

The concept of "atoms" as undivisable units of matter goes back to more than 2000 years ago, and basically the fundamental particles such as quarks and leptons, which we will introduce here, are today's atoms. Historically, however, what was regarded as fundamental in one age turned out to be divisible later, and so the present use of the word "atom" no longer refers to undivisable units, and future centuries might teach us how to split quarks.

Macroscopic matter consists of molecules which are built out of atoms; these atoms define the elements. From the four elements of the ancient Greeks we moved to 92 natural and about 25 artificial elements, each of which may appear in several isotopes. Things became clearer when we learned that each atom consists of a small

© Springer-Verlag GmbH Germany 2017
D. Stauffer et al., *From Newton to Mandelbrot*, Graduate Texts in Physics,
DOI 10.1007/978-3-662-53685-8

nucleus and a large electron cloud around it, with the nucleus again composed of Z charged protons and $A-Z$ neutral neutrons (found by Chadwick in 1932). A neutral atom thus has Z electrons determining its chemical behaviour; thus Z fixes the element, whereas different atomic weights A for the same Z describe isotopes. For example, A is 3 and 4 for the two helium ($Z = 2$) isotopes ^{3}He and ^{4}He, whereas uranium ($Z = 92$) has the natural isotope with $A = 238$ and the splitable isotope with $A = 235$ to make bombs and reactors therefrom. (Uranium was cooked in a cosmic bomb, the supernova explosion.) So, in 1932 we had just three basic particles, the proton, the neutron, and the electron, to build all known tangible matter from.

"Unfortunately" the neutrino ν (=little neutron) was proposed in 1930 by Pauli to take away some of the energy, momentum and spin arising in the beta-decay of a neutron into a proton and an electron. These neutrinos show very little interest in any reactions and have very small mass; on average our bodies react a few times in their life with a neutrino (coming from the sun). Particles of intermediate masses (between that of an electron and that of a proton), called mesons, were found a few years later and called μ and π. Each particle seems to have an antiparticle of the same mass but with a different sign of the electric charge: the positron balances the electron, and the antiproton was found in a particle accelerator built particularly to produce such antiprotons according to $E = mc^2$. The electron and the myon (or heavy electron; originally mislabelled as the μ-meson), together with the quite heavy τ found only in 1975 at the Stanford two-mile accelerator, are called leptons or light-weights; each has its own type of neutrino. For the medium-weight mesons and heavy-weight baryons the proliferation of particle types due to new powerful accelerators was still more drastic; the Δ^{++} baryon carries two electric charges. These artificially produced particles decay after a short time and thus are called resonances: they show up as peaks of finite width$-1/$lifetime in scattering experiments, like in Fig. 3.8. In short, things became very messy in the 1960s, with an ever increasing number of elementary(?) particles.

Simplification came from a mathematical concept of group theory called SU(3) and from the idea of two or three quarks building up the mesons and baryons, as developed by Gell-Mann, Zweig and Ne'eman. Now the proton and neutron are no longer fundamental, but consist of the truly(?) fundamental quarks and the forces between them. Three quarks form a baryon such as the proton, neutron, or Δ, and two quarks form a meson such as π or ρ. (Their name is taken from James Joyce: "Three quarks for Muster Mark". SU(3) is the special unitary group of complex 3×3 matrices U with $\det(U) = 1$ and $U^\dagger = U^{-1}$.) Quarks have electric charges $\pm\frac{1}{3}, \pm\frac{2}{3}$ which explains the existence of the doubly charged Δ^{++} consisting of three quarks with 2/3 charge. Combinations of more than three quarks are also possible.

But again, things are not that simple: instead of two types of quarks they appear in six types (six "flavors"): u, d, c, s, t, b (=up, down, charm, strange, top, bottom). These six flavors are grouped into three generations, which correspond to the three

Table A.1 Fundamental particles of 1994, when evidence for the top quark was found

Leptons			Quarks		
ν_e	ν_μ	ν_τ	u	c	t
e	μ	τ	d	s	b

Table A.2 Masses in GeV and charges in e of fundamental particles

Name	Leptons mass	Charge	Name	Quarks mass	Charge
ν_e	0	0	u	0.005	+2/3
e	0.0005	−1	d	0.010	−1/3
ν_μ	0	0	c	1.5	+2/3
μ	0.105	−1	s	0.2	−1/3
ν_τ	0	0	t	174	+2/3
τ	1.784	−1	b	4.7	−1/3

leptons as shown in Table A.1; all these particles are Fermions obeying the Pauli principle. Thus the mesons with two quarks are Bosons, and the baryons with three quarks are Fermions.

To construct a Δ^{++} from three up quarks of the same charge +2/3 would violate the Pauli prohibition against more than one Fermion in the same state. Thus quarks, in contrast to leptons, appear in three "colors", from where the name quantum *chromo*dynamics comes for quark dynamics. And each of the quarks and leptons has its antiparticle; mesons are formed by one quark and one antiquark. Thus we have at present $6 \times 3 \times 2 = 36$ different quarks of various colors and flavors, and $6 \times 2 = 12$ different leptons, or 48 fundamental particles all together.

The proton consists of two up and one down quark, or in short: $p = (uud)$, the neutron is $n = (udd)$, the π^+ meson is formed by an up quark and a down antiquark, and their weights are 0.938, 0.940, and 0.140 GeV, respectively. (1 GeV = 10^9 electron Volt is an energy unit and, via $E = mc^2$, corresponds to a mass of about $1.6 \, 10^{-24}$ g, roughly the mass of a hydrogen atom.) So normal matter needs only e^-, u, d as constituents, a nice simplification. Table A.2 lists the masses (in GeV) and electric charges (in elementary charges) of the fundamental particles. The masses increase with increasing generation number 1, 2, and 3; the electric charges show a clear sense of order, and the neutrinos all have very little mass: These properties are not completely arbitrary. We also see that some of the masses of the three (up or down) quarks forming the proton or neutron are *much* smaller than the mass of that nucleon: most of the mass is hidden in the interaction energy due to the enormous color forces between quarks.

Experimental investigations of the possible types of reactions show that certain particle numbers are conserved in the sense that the number of incoming particles of this type must equal the number of particles of the same type after the reaction is over. Each of the three lepton generations has its own conserved number of particles, and so do the quarks for all generations together. Also the electric charge is always conserved, whereas the mass can be transformed into energy and back. However, antiparticles always count negative for the particle number and charge (not for mass). Thus radiation energy can form an electron-positron pair since then the number of e-leptons is still zero.

With these conservation laws we now understand a crucial difference between mesons (quark + antiquark) and baryons (three quarks): the meson number is not conserved since the quark and antiquark can annihilate each other; the baryon number is conserved since their quark number, whatever combination of antiquarks and quarks we try, is never zero. Indeed, the neutral pion π^0 decays into radiation within a microsecond whereas the proton is stable. The free neutron decays after 15 minutes into a proton, an electron, and an anti-electron-neutrino but this is allowed since both the proton and the neutron have the same baryon number of one. (A neutron star or pulsar does not decay into protons because of the strong forces between neutrons.)

Summary: baryon = three quarks; meson = quark and antiquark. Leptons do not consist of quarks. For quarks and for each of the three lepton generations the number is conserved, with antiparticles counting negative. Also electric charge is conserved, with antiparticles having the opposite charge.

So, why do we not observe in nature those fractional electric charges like 1/3 and 2/3? It seems the force between two quarks is quite strong and for large distances about 0.14 MegaNewton, independent of distance. So if we try to pull them apart, we need so much energy that we merely create new particles: somewhat like north and south poles of a magnetic dipole, we cannot observe quarks isolated. Only "white" combinations of quarks, where the color forces have cancelled each other (like quark-antiquark, or three quarks with the three fundamental colors), are observed as isolated particles. Now we look at the mechanism of these and other forces.

A.1.2 Forces

Gravitation and electric forces have already been introduced in the first two chapters; for two protons in an atomic nucleus, gravitation is about 10^{36} times weaker than their Coulomb force. (We still feel gravitation since there are no negative masses, in contrast to positive and negative electric charges which cancel each other in their force over long distances.) As Fig. 2.4 indicates, electric forces do not propagate with infinite velocity but only with the large but finite light velocity c. Light waves are called photons (γ) in quantum theory, that is Coulomb forces are transmitted via the quasi-particles called photons. Similarly, gravitational forces propagate with velocity c with the help of quantized gravity waves called gravitons (not yet detected as quantized quasi-particles). Photons (and gravitons) follow Bose statistics. Quite generally, forces are supposed to come from the exchange of intermediate Bosons. Decades ago we were taught that exchange of pions creates the "strong" or nuclear force which keeps the protons in a nucleus together against their ten times weaker

Coulomb repulsion. Now pions are no longer fundamental, this example is no longer valid, but the principle remains: forces arise from Boson exchange.

A simple analogy with this type of force are two grumpy old men playing tennis on a frozen lake: the small tennis balls give their momentum to each of the men when they are thrown back, and then the two players are slightly repelled from each other. If they used boomerangs, they would be attracted instead; see the movies for other types of intermediate particles.

The color forces between quarks are transmitted by gluons (i.e., by particles glueing the quarks together) of zero mass. They bind three quarks together as a nucleon (proton or neutron). At some distance from this nucleon some remnant of the color forces is felt, since they have not canceled each other exactly; this remnant is the strong force. Similarly, outside of an argon atom the positive and negative electronic charges cancel each other mostly but not fully; the remnants are the Lennard-Jones forces of Sect. 1.1.3 e. But neither these Lennard-Jones forces (or the exchange energy of homopolar binding, Sect. 3.3.6) nor the strong force between nucleons are fundamental. Nevertheless, these strong forces are 20 times stronger than the electromagnetic force for two protons in a nucleus.

The beta-decay of a proton (uud) into a neutron (udd) plus leptons does not involve color forces, since from the quark point of view an up is merely transformed into a down, a member of the same generation. Therefore this decay within a quarter of an hour is much slower than the color-force decay of a hadronic resonance within typically 10^{-22} s. The responsible force is called the "weak" force and is induced by the exchange of the weak $W^{\pm}$ and Z^0-Bosons, detected in 1982 at CERN, with masses of 81 and 91 GeV, respectively. For protons in a nucleus, their weak force is about ten million times weaker than their electric repulsion.

Summary: Forces are based on virtual Boson exchange: gluons of zero mass for the color force, photons of zero mass for the Coulomb force, and quite heavy weak Bosons for the weak force.

These intermediate Bosons (virtual particles) are packets of energy $\Delta E = mc^2$ with a short lifetime Δt, such that the energy-time uncertainty relation $\Delta E \, \Delta t \simeq \hbar$ allows their creation. If they move with a velocity close to c, within their lifetime they cover a distance of the order of the Compton wavelength:

$$\Delta r = c\Delta t \simeq c\hbar/\Delta E = \hbar/mc \, .$$

The heavier a particle is the less it can run and the shorter is the distance over which the interactions are felt. The force law then contains a factor $\exp(-r/\Delta r) = \exp(-rmc/\hbar)$ apart from pre-exponential factors. For a pion, the resulting Δr is of the order of a proton radius, 10^{-12} cm, as it should be if the "strong" proton-proton interaction came from pion exchange. Coulomb forces and gravitation are felt over infinite distances without exponential cut-off and thus have zero mass. Color forces also must have infinite range since otherwise we could isolate single quarks; thus also the gluons are massless. The weak interaction covers only very short distances because of the high mass of the corresponding intermediate Bosons.

Why are the $W^\pm$ and Z^0 so heavy? You can blame the US Congress. The so-called Higgs particles are connected with breaking a spontaneous symmetry. This is an effect already seen in Ising models: the interaction energy remains the same if all spins reverse their orientation; thus the Hamiltonian is up-down symmetric. The paramagnet at $T > T_c$ also has up-down symmetry. But if we cool it down to $T < T_c$ the actual state of the system loses this symmetry: either the majority of spins are up, or the majority of spins are down. This broken symmetry leads to domain walls or (in isotropic magnets) to spin waves or magnons. Similar broken symmetries are supposed to cause the mass of the (virtual) particles. Experimental evidence would be the detection of the Higgs particle, for which purpose construction of the superconducting super collider SSC was started in Texas. Then the project was stopped by parliament as being too expensive. Gell-Mann made a nice comment as to why fellow Nobel laureate Phil Anderson did not support SSC. The European Large Hadron Collider in 2012 filled the gap and found it at a mass of 126 GeV. Thus if the first author's Body Mass Index is too high, blame the Nobel laureates (2013) Higgs and Englert for this obesity.

What has been described here is the so-called *standard model* which includes color forces. The Grand Unified Theory (GUT) combines it with electromagnetic and weak forces, and the Theory of Everything would include gravity. If that theory could be established, it would be a good basis for starting anew the search for the true fundamentals, as has happened in the past: Max Planck was warned against taking up physics as a student, by being told that nearly all of physics is understood.

A.2 Quantum Field Theory of Elementary Particles

In magnetism, we understand on the basis of suitable assumptions, such as the Ising models, how a spontaneous magnetization can be formed. And we learned how to simulate this Ising model on a computer. Analogous computer simulations for quantum chromodynamics (QCD), i.e., for the forces building up the nucleons out of quarks, today constitute one of the most hungry consumers of computer time. Even special-purpose computers were built, which can simulate only QCD, but do this particularly efficient. These methods and their underlying theories are too complicated for our purpose; thus we only give here an impression of how it is done and what the results are.

A.2.1 Quantum and Thermal Fluctuations

What is the average energy of an harmonic oscillator? If we could have neither quantum nor thermal fluctuations, the oscillator would stay forever in its potential minimum of zero energy E. Both types of fluctuations are included in (4.48):

$$E = \hbar\omega(\langle n\rangle + 1/2), \quad \langle n\rangle = 1/[\exp(\hbar\omega/kT) - 1], \tag{A.1}$$

In the limits of very high and very low temperatures this energy becomes

$$E = \frac{\hbar\omega}{2}(\hbar\omega \gg kT), \quad E = kT(\hbar\omega \ll kT) \tag{A.2}$$

corresponding to zero-point motion and equipartition theorem, respectively. Thus we see that $\hbar\omega$ plays, at low temperatures, the role which at high temperatures is played by kT. Similarly, we saw in the WKB approximation for tunnel effects that the wave function "propagates" as $\exp(-S/\hbar)$ through an energy barrier at zero temperature; at finite T without quantum effects an energy barrier would be overcome with probability $\exp(-E/kT)$: Again we found a correspondence of Boltzmann's constant (thermal fluctuation) with Planck's constant (quantum fluctuation). In this way, also the Metropolis simulations of the Ising model, Sect. 4.3.9, have a counterpart in *quantum field theory* at zero temperature.

Such theories generally start from Feynman's path integral formula

$$Z_s = \sum e^{iS/\hbar} \tag{A.3}$$

for the partition function, instead of the old thermal partition function

$$Z = \sum e^{-E_n/kT} \tag{A.4}$$

of (4.2). Now the "action" S plays the role of Hamiltonian H or energy and is defined through the Lagrange function L:

$$S = \int L\, dt. \tag{A.5}$$

The integral in (A.5) runs from one fixed point in time to another fixed point in time, just as for Hamilton's principle in (1.32). (We set the velocity of light equal to unity to have the same units for time and length.) And the sum in (A.3) runs over all possible paths $r(t)$ from one fixed point to the other. To sum the thermal partition function (A.4), we learned to use a finite volume V and to replace summations over all wavevectors Q by integrals $(V/2\pi)^3 \int d^3Q$. To distinguish between different paths for (A.3) we regard space and time as discrete variables, incremented by a minimum length a and a minimum time $\Delta\tau$ instead of varied continuously. The paths in (A.3) are thus time-ordered connected chains of bonds in a four-dimensional hypercubic space-time lattice. We see this easier if we restrict ourselves to one space dimension x where the path is a function $x = x(t)$ of time t. In the discretized form, this path is a chain of bonds on a square lattice with time (in units of $\Delta\tau$) plotted horizontally and space (in units of a) plotted vertically.

In this one-dimensional case, the Lagrange function L for a single particle in a potential $U(x)$ and with kinetic energy $mv^2/2$ is $L = \frac{m}{2}(dx/dt)^2 - U$. With the imaginary (so-called Euclidean) time $\tau = it$ and the Euclidean action $S_E = -iS$ we get

$$L = -\frac{m}{2} (dx/d\tau)^2 - U = -H ,$$

with the usual Hamiltonian or energy; also

$$S_E = -iS = -\int L \, d\tau = \int H \, d\tau ,$$

and thus

$$Z_s = \sum e^{-S_E/\hbar} = \sum \exp \left(-\int E(\tau) \, d\tau/\hbar \right) = \sum \prod_n \exp \left(-E_n \Delta\tau/\hbar \right) \quad (A.6)$$

looks very much like the thermal partition function (A.4) since we replaced the integral over τ by the discrete sum imposed by our lattice approximation. We just have $\hbar/t(N_t \Delta\tau)$ instead of kT. (The time integration over τ goes from 0 to $t = N_t \Delta\tau$, which is the analog of $\hbar/kT$: the larger T, the smaller is the length N_t in the time direction. Our discretization then transforms this time integral into a sum over N_t terms.)

A.2.2 Simulations at $T = 0$

Now the principle of a Monte Carlo simulation can be taken from the Metropolis program for Ising models:

1. choose a lattice site;
2. change slightly the variable belonging to this site;
3. calculate the resulting action change ΔS_E;
4. calculate a random number z between 0 and 1;
5. accept the change if $z < \exp(-\Delta S_E/\hbar)$;
6. calculate the required quantities;
7. choose a new site and start again.

Our lattices are four-dimensional, of size $L \times L \times L \times N_t$, for three space ($L$ sites) dimensions and one time (N_t sites) dimension. We obtain not the absolute masses of elementary particles but the ratios of their masses, just like in an Ising model we do not calculate the absolute value of the magnetization (in electromagnetic units per cubic centimeter) but its ratio to the saturation magnetization, as a function of the ratio J/kT. The only problem is now that the variable at each lattice site is no longer an Ising spin represented by $+1$ or -1 but a complex 3×3 matrix arising from the mysteries of SU(3) group theory (fundamental three-dimensional adjoint representation). And thus the simulated lattices of quantum chromodynamics are much smaller than the 336^4 sites of simulated Ising models.

Nevertheless, with lattices of sizes up to $30 \times 32 \times 32 \times 40$ and careful extrapolations to infinite lattices, the group of Weingarten at IBM obtained a quarter-century ago mass ratios for many hadrons which agreed within a few percent with the experimentally observed ratios. They built their own parallel computer, the GF11 (its name comes from 11 GigaFlops, i.e., 11×10^9 floating point operations per second) and used it for about one year for these calculations.

A.2.3 Simulations in the TeraKelvin Region

Just as with the harmonic oscillator, for QCD we can look at both quantum and thermal fluctuations together. At room temperature, the quarks will hardly feel this temperature, since one electron Volt (1 eV) corresponds to 10^4 K, and important quark energies are of the order of GeV. Thus only at temperatures of a TeraKelvin $= 10^{12}$ K are new effects found, such as the quark-gluon plasma.

For water we know that for temperatures below the boiling temperature (or the critical temperature), water droplets can be formed out of supersaturated vapor; if we increase the temperature beyond this limit, the droplets evaporate into vapor molecules which may form a nearly ideal gas. Similarly, at low temperatures the quarks form droplets of three quarks each, such as protons and neutrons which together with electrons, may form atoms, molecules, and life. For temperatures of one TeraKelvin, these nucleons "evaporate" into their three quarks; the gluons can no longer keep the quarks together. However, quarks at a long distance, in contrast to water molecules, still feel a strong force of the order of the weight of 16 tons, and thus they do not form an ideal gas. Instead, together with the gluons, they form a quark-gluon plasma, a soup where no long-lived bonds exist between these quarks. The transition temperature is about 2×10^{12} K corresponding to 0.2 GeV, of the order of the pion mass. It seems we have here a first-order transition (at least for quenched QCD) where the order parameter jumps to zero (like boiling water under atmospheric pressure, and not like the Ising Curie point). In one such simulation 20,000 Cray hours were invested.

Water has an unusually high surface tension at room temperature. Also the transition to a quark-gluon plasma seems to be associated with a surface tension of $5 \, \text{MeV/(Fermi)}^2$, which is roughly 10^{19} times larger than that for water, but rather small by quark standards (1 Fermi $= 10^{-13}$ cm $=$ typical nucleon size). Shall we go swimming in it?

You may wonder how in a quenched approximation without moving quarks one can form nucleons at low temperatures out of the quark-gluon plasma. One cannot, but one sees something nevertheless: the gluons alone are thought to form glueballs since gluons, unlike the photons of electromagnetism, interact with each other. (QCD, in contrast to the Maxwell equations, is an intrinsically *nonlinear* theory.) On a more formal level, a phase transition may be connected with a broken symmetry: the interaction energy of the Ising model does not change if all down spins are flipped up and all up spins are flipped down. Thus the Ising Hamiltonian has up-down symmetry. However, for $T < T_c$ the Ising magnet has a spontaneous magnetization which favors one of the two orientations: The up-down symmetry is broken, the cooling of the quark-gluon plasma may similarly lead to a broken "chiral" symmetry.

All these theoretical studies would have been much easier had we watched out better shortly after the big bang of the present universe, some 10^{10} years ago. At that time, room temperature was indeed two Tera-Kelvin and more (if rooms were available at that time). Today we have to recreate such high temperatures with lots of effort in powerful particle accelerators. Perhaps future studies will show new phases of quark matter, or new and heavier quarks as part of an infinite spectrum of resonances, of which we have just seen the lower end with the six quark flavors of up, down, charm, strange, top, and bottom.

Questions

1. What did all natural matter consist of: (a) 2000 years ago; (b) 125 years ago; (c) 90 years ago; (d) for the last 50 years; (e) in 100 years time?
2. How many quarks form a meson, and how many form a baryon?
3. For which particle numbers do conservation laws hold?
4. What are baryons? What are μ-mesons?
5. Which fundamental forces are known, and what is their ratio?
6. What is the Theory of Everything supposed to do?
7. Which forces are transmitted by which particles?
8. What is the partition function of quantum field theory, analogous to $Z = \sum_n e^{-\beta E_n}$?
9. How do quantum fluctuations at zero temperature correspond to thermal fluctuations at high temperature, for the example of the harmonic oscillator?
10. How heavy are Cupid's arrows? ($\Delta r = 10\,\text{m}$) (Cupid = Amor $\simeq$ Eros = ancient god of a dangerous disease; according to Einstein, gravity is not responsible for people falling in love.)

Appendix B
Answers to Questions

These answers should not be looked up in order to learn them, but to check if the reader has learned enough from the text. They are particularly suited to the preparation for an oral examination. Other answers may be also be reasonable; interpret questions such that an answer can be given.

B.1 Mechanics

Sect. 1.1

1. The orbital period is proportional to (major semi-axis)$^{3/2}$ for all solar planets.
2. If measured in an inertial system of coordinates.
3. $f = m\omega^2 r = mv^2/r$.
4. $mv^2/2 = GMm/R$, with the Earth's radius R and mass M; stone mass m cancels.
5. $GM/R = g = 10^3 \text{ cm/s}^2$ with the Earth's radius R and mass $M = \rho(4\pi/3)R^3$.

Sect. 1.2

6. $\mu = m_1 m_2/(m_1 + m_2)$.
7. $\sum_i (F_i - m_i d^2 r_i/dt^2)\delta r_i = 0$.
8. $\sum_i F_i \delta r_i = 0$ or $\delta U = 0$.

Sect. 1.3

9. It would give nonsense if we moved the end point into the starting point.
10. Lagrange: coordinates and velocities. Hamilton: coordinates and momenta.
11. Atomic vibration; for zero wavevector, the acoustic (optic) phonon frequency is (non)zero.

© Springer-Verlag GmbH Germany 2017
D. Stauffer et al., *From Newton to Mandelbrot*, Graduate Texts in Physics,
DOI 10.1007/978-3-662-53685-8

Sect. 1.4

12. $M = dL/dt$; $L = \Theta \omega$; ω is only an axial vector.
13. If you put coordinate axes into the principal axes, the inertia tensor Θ is diagonal; Θ has as the diagonal elements the moments of inertia about each axis.
14. Zero, because a cube has three equal principal moments of inertia, like a sphere.
15. No, they are linear in ω_1 and ω_2: amplitude cancels out.
16. Because of the cross product: $dL/dt = r \times f$.
17. $\omega \sim$ torque for symmetric gyroscope; used for NMR (magnetic resonance imaging).

Sect. 1.5

18. For ∂ the coordinate is fixed, for d it moves with the masses.
19. $\partial \rho / \partial t = \operatorname{div} j$.
20. $\sigma = C\varepsilon$; in isotropic fluids: $\sigma_{ik} = -P\delta_{ik}$.
21. Hurricane (over Atlantic Ocean) = typhoon (over Pacific Ocean); a tornado is much smaller (over some continents).
22. $d\rho/dt = 0$; $\operatorname{curl} v = 0$; zero viscosity; $dv/dt = 0$; $v = 0$.
23. For small values only.
24. $D = k_B T / 6\pi \eta R$.

B.2 Electricity and Magnetism

Sect. 2.1

1. $U = eq/r$; $f = eq/r^2$.
2. $\operatorname{div} E = 4\pi \rho$; $\operatorname{curl} E = 0$; $\operatorname{div} B = 0$; $\operatorname{curl} B = 4\pi j$.
3. Lorentz force $= qv \times B/c$.
4. $u = (E^2 + B^2)/8\pi$; $S = (c/4\pi)E \times B$.
5. How to build up an arbitrary function out of superposition of sinus waves.
6. Potential = integral over density/distance, with ρ at time $t - R/c$. The electric forces need a time R/c to reach a distance R.
7. G is the effect (at time t) of a cause of unit strength at the origin (at time 0).
8. Put together a positive and a negative charge at a very small distance.
9. Zero; only the torque is nonzero. A field gradient would give a force.

Sect. 2.2

10. Potential = charge/$(r - rv/c)$.
11. Not really; we merely approximate the atomic structure through M and P.

Sect. 2.3

12. Look at the moon with a flashlight and move hand back and forth.
13. By a factor $1/(1 - v^2/c^2)^{1/2} \simeq 1 + v^2/2c^2$.
14. Not at all; but one can write them nicer in four dimensions.
15. The four-momentum $(\boldsymbol{p}, icm)$ transforms like $(\boldsymbol{r}, ict)$.

B.3 Quantum Mechanics

Sect. 3.1

1. Photo effect, stability of atoms, electron microscope, uncertainty of next exam?
2. No, only if the operators commute.
3. The unit operator. $|n\rangle\langle n|$ is an operator, $\langle n|n\rangle = 1$ a scalar product.
4. It must be unity since it is the space integral over the probability density.

Sect. 3.2

5. $-(\hbar^2/2m)\nabla^2\Psi + U(\boldsymbol{r})\Psi = E\Psi \; (= i\hbar\partial\Psi/\partial t)$.
6. If U is finite, then Ψ'' is finite and thus Ψ' is continuous.

Sect. 3.3

7. $Z = 9$ and 10: first and second level; $Z = 11$ also one electron in third level.
8. Photons coming from jumps to the first, second, third level, from higher levels.
9. Both have two protons, helium 3 has one neutron (Fermion), helium 4 has two (Boson).
10. Coulomb plus quantum mechanical indistinguishability gives exchange energy.

Sect. 3.4

11. To the square of the matrix element v_{kn}.
12. Elastic scattering gives Fourier transform in space, inelastic one gives it also in time.
13. Frequency dependence $\sim 1/(1 + \omega^2\tau^2)$.

B.4 Statistical Physics

Sect. 4.1

1. $\frac{1}{4} 10^{20}$.

2. $2^N \simeq 10^{0.3N}$ configurations, $N = L^3 = 1, 8, 27, 64$; $1\,\mu s$ per configuration.

Sect. 4.2

3. $dE = T dS - P dV + \mu dN + \dots$ Extensive variables as differentials.
4. $E' = E -$ (intensive variable λ) $\times$ (extensive variable Λ) has λ as natural variable.
5. Helmholtz free energy F and $F + PV - BM$.
6. $C_V = T(\partial S/\partial T)_V$; $C_P = T(\partial S/\partial T)_P$; $\chi_M = (\partial M/\partial B)_M = 0$ (nonsense).
7. Vapor pressure and latent heat of vaporization.
8. $P_{osm} V/kT =$ number of dissolved molecules.

Sect. 4.3

9. $\langle n \rangle = 1/(e^{(\varepsilon - \mu)/kT} \pm 1)$ (Fermi: $+1$. Bose: -1. Maxwell: ± 1 negligible).
10. $S/kN = \log(\text{const.} E^{3/2} V/N^{5/2})$; const. depends on $\hbar$!
11. Linear in T.
12. $\varepsilon_F = kT_F = p_F^2/2m = \mu$ for Fermi energy ε_F, Fermi temperature T_F, Fermi momentum p_F.
13. Proportional to $T^{d/b}$ in d dimensions, if $\omega(Q \to 0) \sim Q^b$.
14. With third power for acoustic and infinite power for optic phonons. (Infinite power means $\exp(-\hbar\omega/kT)$; try to derive this: Debye versus Einstein.)
15. With $T^{3/2}$ for isotropic ferromagnets and T^3 for antiferromagnets.
16. A Taylor expansion in the density, to approximate real gases.
17. The line for liquid-vapor equilibrium cuts off equal areas in the P–V diagram.
18. P/P_c is a material-independent function of T/T_c and V/V_c.
19. $m = \tanh(\beta B + m T_c/T)$; $T_c = 0$ without interactions.
20. With the exponent $\delta = 3$ in old theories and nearly 5 in reality.

B.5 Appendix A: Elementary Particles

1. (a) fire, water, air and earth; (b) nearly 100 elements; (c) proton, neutron, electron; (d) quarks, leptons, ...; (e) heavy Staufferons, medium Stanleyons, and light Lesneons?
2. Quark and antiquark for mesons, three quarks for baryons.
3. For the quarks and for each of the three lepton generations.
4. Baryons are proton, neutron, and higher Fermi resonances. μ-mesons = wrong name for myons (leptons); like walfish.
5. Gravitational, weak, electromagnetic, and strong or color interactions. Ratio $10^{-36} : 10^{-7} : 1 : 10^1$ for two protons in a nucleus.
6. Should describe all four forces between all particles.
7. Electromagnetic forces by photons, strong forces by gluons, weak forces by $W^\pm, Z^0$.
8. $Z_s = \sum e^{iS/\hbar}$ with the action S.
9. Energy $= \hbar\omega/2$ at $T = 0$ corresponds to Energy $= kT$ for $T \to \infty$.
10. $m = \hbar/c\Delta r = 10^{-27-10.5-3}$ g. Do you want to rely on *that*?

Index

A

Acceleration, 2
Amplitude, 191
Anagni, 191
Angular momentum, 9, 110
Angular velocity, 4
Ant, 193
Areal force, 47
Asymptotic, 218
Atkins, 161
Atomic structure, 93, 115
Attractor, 232, 237
Atwood, 23
Autonomous, 216

B

Balmer series, 115
Basin of attraction, 226
Bernoulli, 51
Bifurcation, 230, 233, 237
Binding, 118
Biot-Savart, 66
Birkhoff ergodic theorem, 242
Boltzmann, 129, 243
Born, 124
Bose, 117
Bound state, 106, 228
Broglie, Louis de, 99

C

Canonical equations, 27, 219
Canonical probability distribution, 140
Capacitance, 82
Catastrophe theory, 235
Central force, 10

Centre of mass, 18
Chaos, 221, 232, 240
Characteristic length, 194
Chemical kinetics, 220
Clausius-Clapeyron equation, 146
Clausius-Mossotti formula, 82
Cluster, 168
Commutator, 27
Condensation, Bose-Einstein, 158, 206
Condenser, 81
Conservation laws, 4, 18
Conservative, 6, 232
Constraining force, 21
Continuity, 46
Continuum, 42
Contraction, 86
Control parameter, 233
Coriolis, 5
Correspondence principle, 103
Coulomb potential, 112
Coulomb's law, 61
Critical dimension, 201
Critical point, 144, 163, 166
Cross-product, 10
Crossover time, 199
Curie temperature, 77
Curie-Weiss law, 166
Curl, 43
Cycle, 231, 237

D

D'Alembert, 22
Data collapsing, 200
Debye, 161
Degeneracy, 115
Delta function, 63

© Springer-Verlag GmbH Germany 2017
D. Stauffer et al., *From Newton to Mandelbrot*, Graduate Texts in Physics,
DOI 10.1007/978-3-662-53685-8

Derivative, total, 44
Diagonalization, 223
Dielectric, 76
Diffusion limited aggregation, 207, 213
Dipole, 65, 72–74
Dirac, 95
Dissipative, 233
Divergence, 43
Dynamical system, 216

E
Eigenfunction, 111
Eigenvalue, 30, 95, 223
Einstein, 83
Elasticity, 42
Electron microscope, 94, 105
Ellipse, 2
Energy, 6
Energy density, electrical, 68
Entropy, 132
Equilibrium, 132, 221
Equipartition law, 155
Ergodicity, 242
EULER, 38
Euler notation, 45
Euler's equations, 38
Exchange interaction, 119

F
Fermi-gas, 156
Fermi particles, 117
Ferromagnet, 77
Field strength, 62
Field tensor, 86
Fixed point, 221
Flow, 217
Force, 3
Four-vector, 87
Fourier, 69
Fractal, 56, 168, 189
Fracture, 213
Frustration, 120
Functional equation, 196

G
Galileo, 2, 4
Gauge, 67
Gauss, 46, 63
Gelation, 205
Generalised coordinates and momentum, 24, 26
Gibbs, 137

Golden Rule, 123
Gravitation, 11
Gravitational acceleration, 2
Green, 71
Growth sites, 207
Gyroscope, 38

H
Hamilton, 25, 26
Hamiltonian, 219
Harmonic approximation, 28, 229
Harmonic oscillator, 13, 107, 250
Hartman-Grossman theorem, 222
Heat pump, 143
Heisenberg, 100
Helmholtz, 137
Henry, 147
Hermite, 95, 109
Hilbert, 95
Hooke, 48
Hopf bifurcation, 234, 237, 246
Horoscope, 41
Hydrodynamics, 42, 50
Hydrogen atom, 112

I
Ideal gas, classical, 151
Incompressible, 47
Indistinguishability, 116
Induction, 66, 80
Inertial system, 3
Inertia tensor, 34
Infinite cluster, 205
Inhomogeneous, 70
Integrable system, 219
Intermittency, 245
Invariant measure, 242
Inversion temperature, 150
Ising, 78, 131
Ising model, 78, 131, 164, 203

K
KAM theorem, 246
Kepler, 2

L
Lagrange equations, 21, 24
Lagrange function, 24
Lagrange notation, 45
Landau, 166, 246
Laplace, 51, 63, 64

Larmor, 41
Latent heat, 145
Lattice animals, 204
Legendre polynomials, 111
Legendre transformation, 137
Length contraction, 86
Lennard-Jones, 11
Lienard-Wiechert, 75
Light, 61, 68
Linearization theorem, 222
Linearized evolution, 222
Liouville theorem, 219, 232
Logistic map, 236
Lorentz transformation, 83
Lorenz equations, 220
Lyapunov exponent, 239
Lyman series, 115

M
Magnetic field, 65
Magnon, 109, 161
Mandelbrot, 56, 189
Mass, 3
Mass action law, 146
Matter, 75
Maxwell construction, 163
Maxwell distribution, 151
Maxwell equation, 66
Mean field theory, 164
Metro, 203, 211
METROPOLIS, 168
Michelson-Morley, 83
Mode, 246
Model, 216
Momentum, 4
Monopole, 65
Monte Carlo, 78
Multifractal, 209
Multipole, 74
Myonen, 85

N
Navier-Stokes, 54
Neuron, 211
Newton, 3, 218
Node, 224
Noether, 26
Normal form, 235
Normal vibrations, 29
Nucleation, 145
Nutation, 39

N-vector model, 203

O
Operator, 95, 98

P
Paramagnet, 77
Partition function, 130
Paschen series, 115
Pauli, 117
Pendulum, 8, 221, 250
Penetration, 102
Period-doubling bifurcation, 237, 243
Periodic system, 115
Permeability, 77
Perturbation, 120
Phase equilibrium, 143
Phase portrait, 217
Phase space, 216
Phase transition, 234
Phonon, 31, 109
Photoeffect, 94, 122
Photon, 109
Pitchfork bifurcation, 234, 237, 243
Planet, 13
Poincaré section, 237
Poiseuille, 55
Poisson bracket, 27
Poisson ratio, 49
Polar coordinates, 5
Polarisability, 82
Polarisation, 76
Polymer, 201
Position vector, 1
Potential, 6, 227
Potential flow, 51
Potts model, 203
Power, 6
Poynting, 68
Precession, 34, 40
Principal axes, 35
Principal quantum number, 114
Probability, 96, 129

Q
Quantity of heat, 136
Quasi-periodic, 232
Quasiparticle, 109

R

Raoult, 148
Rayleigh, 75
Relativity, 82
Renormalization group, 205
Restrictive conditions, 20
Reynolds, 56
Rigid bodies, 33
Route to chaos, 243
Ruelle-Takens route, 246

S

Saddle, 224
Saddle-node bifurcation, 233, 237, 245
Scalar product, 6
Scaling theory, 169
Scattering, 124
Schrödinger equation, 101
Schrödinger paradox, 97
Self-avoiding walk, 202
Self-catalytic reaction, 220
Separatrix, 226
Sierpinski gasket, 190
Similarity, 56
Snowflakes, 189, 214
Specific heat, 132
Spherical harmonic, 111
Spherical model, 206
Spin, 115
Spiral point, 224
Spontaneous magnetisation, 77
Stability, 221
Stability matrix, 222
Stable manifold, 227
Stefan-Boltzmann, 161
Stokes, 51
Strain tensor, 43
Strange attractor, 241, 246
Stress tensor, 47
Subharmonic cascade, 245
Superconductivity, 159
Surface, 79
Susceptibility, 77
Symmetry, 225

T

Tensor, 35

Thermodynamic relation, 138
Thom, 235
Time dilatation, 86
Torque, 9
Trace, 44
Trajectory, 216
Transcritical bifurcation, 234
Triple product, 10
Tunnel, 104
Turbulence, 213
Two-body problem, 19

U

Uncertainty, 94, 100
Universality, 167, 243
Universality class, 200, 245
Unstable manifold, 227

V

Van der Waals, 162
Van't Hoff, 149
Vapour pressure, 144
Vector, 33
Vector potential, 67
Velocity, 2
Virial expansion, 161
Virtual displacement, 22
Viscosity, 53
Viscous fingers, 213
Voltage, 63
Volume force, 47
Vortex, 51

W

Water, 213
Waves, 49, 68
WKB approximation, 105

Y

Yukawa, 11

Z

Zeeman, 113

Printed in the United States
By Bookmasters